Lecture Notes in Morphogenesis

Lecture Notes in Morphogenesis is an interdisciplinary book series which aims to face the questions of emergence, individuation and becoming of forms from several different points of view: those of pure and applied mathematics, of computational algorithms, of biology, of neurophysiology, of cognitive and social structures. The set of questions above concerns all the manifestations of Being, all the manifestations of Life. At the heart of contemporary embryogenesis lies an essential question: How can form emerge from the constant, chaotic flow? How can a sequence of purely informational elements—an a-signifying combination of chemical substances organized in the DNA molecule—evolve into the highly complex and structured forms of the living organism? A similar question can be asked when we deal with the morphogenesis of vision in neural systems and with the creation of evolving synthetic images, since digital technology makes possible the simulation of emergent processes both of living bodies and of visual forms. Finally the very idea that abstract structures of meaning could be captured in terms of morphodynamic evolution opens the door to new models of semiolinguistics, semiotic morphodynamics, and cognitive grammars. An entire heritage of ideas and concepts has to be reconsidered in order to face new and challenging problems: the theoretical framework opened by Goethe with the introduction of the word "Morphogenesis" is developed by D'Arcy Thompson in "On Growth and Form", it is reorganized with new theoretical insights by the classical structuralism of Levi-Strauss and formalized by the dynamical structuralism of René Thom. The introduction of the post-structuralists ideas of individuation (in Gilbert Simondon and Gilles Deleuze) and plasticity of structures builds a bridge to contemporary problems of morphogenesis at a physical, biological, social and transindividual level. The objective of this book series is to provide suitable theoretical and practical tools for describing evolutionary phenomena at the level of Free boundary problems in Mathematics, Embryogenesis, Image Evolution in Visual Perception, Visual Models of Morphogenesis, Neuromathematics, Autonomy and Self-Organization, Morphogenetic Emergence and Individuation, Theoretical Biology, Cognitive Morphodynamics, Cities Evolution, Semiotics, Subjectivation processes, Social movements as well as new frontiers of Aesthetics. To submit a proposal or request further information, please use the PDF Proposal Form or contact directly: *Dr. Thomas Ditzinger (thomas.ditzinger@springer.com)*

Sabine Ploux

Semantic Matter and Its States

Words, Meaning, and Modeling

Translated from French by Vivian Waltz

Springer

Sabine Ploux
CAMS EHEES-CNRS
Paris, France

ISSN 2195-1934 ISSN 2195-1942 (electronic)
Lecture Notes in Morphogenesis
ISBN 978-3-032-22808-6 ISBN 978-3-032-22809-3 (eBook)
https://doi.org/10.1007/978-3-032-22809-3

This Springer imprint is published by the registered company Springer Nature Switzerland AG
The registered company address is: Gewerbestrasse 11, 6330 Cham, Switzerland

Competing Interests The author has no competing interests to declare that are relevant to the content of this manuscript.

Introduction

Lexical semantics has long been studied using philosophical, linguistic, and psychological approaches. Starting in the nineteen fifties, in line with a series of previous works [1, 2], attempts to formalize this domain led to the proposal of mathematical models based on logic or tree diagrams that applied derivation rules (Noam Chomsky [3], Richard Montague [4], etc.). In an alternative continuist approach, data analyses based on matrix algebra (Jean-Paul Benzécri [5]) or dynamic systems (René Thom [6]; Jean Petitot [7]) were also put to work. These different kinds of models served as a framework for computer-based designs, including ontological trees representing the structure of the links between lexical concepts [8], matrix reduction algorithms used to implement analyses of textual data (such as LSA [9] and Semantic Atlases [10]), and propagation algorithms based on artificial neural networks to depict how lexical processes converge to form concepts or semantic values. All of these processes can be seen as dynamic systems [11].

Current vector-based neural networks (NNs) as models of usage. Recently, this domain has benefitted from a surge of interest and development resulting from a new way of using the paradigm of artificial neural networks (NN) (see for example some of the earliest works [12, 13]). This surge was made possible not only by the increased size of available digital text corpora and the greater computing power of computers, but also by a change in approach as to how to use neural networks for language. The method was initially founded on learning the word-distribution probabilities in sentences or *n*-word windows. The question of what is the best network architecture is a prolific domain, with *transformers* constituting the dominant choice [14]. Although dynamically implemented, these NNs propose the vector space as the mathematical paradigm underlying the representation of language units. And, in the end, every word, phrase, or sentence assigned a vector composed of the values of the learned neurons' weights. The results obtained by these models are massive and provide a representation of the set of all units contained in the lexicon, whereas the older methods—also based on artificial neural networks—simply attempted to reproduce some of the highly fragmented processes brought to the fore by linguists or psycholinguists.

If a vector embedded in a space is associated to a word, its semantics is characterized by the set of all proximities or differences that distinguish it from other word vectors. Earlier studies on data analysis had already been rooted in this same vector paradigm. Moreover, matrix reduction—at the heart of data analysis methods—is still used by neural networks to decrease the huge size of the spaces they generate. However, over and above this similarity in representation format, neural networks can integrate the sequential order of the appearance of words in the very sentence implemented in the ordered neural structure and the neural layers, which is something that linear and matrix-based methods cannot do.[1]

Advantages of the neural paradigm. The current approach consists of computing large generic vector spaces in a preliminary phase, and then proceeding, in the subsequent phases, to the learning of more specific linguistic tasks or automatic language processing. These dedicated models included as input the vectors learned in the first phase. They are thus successfully used in automatic translation and in question-answer synthesizing systems, etc. The vector spaces learned also act as parameters for studying neurolinguistic processes. They guarantee a more accurate resolution of the cortical areas involved in the semantic processing of words than was possible with earlier protocols based on the manual choice of word categories (for a forerunning study, see [16], or even [17]).

Limitations of the above paradigm. However, beyond these technical achievements, which, again, rely on resources and colossal computations in more recent artificial neural networks, the vector paradigm, in which every word is assigned a vector, failed to account for linguistic properties and processes like polysemy[2] (the variation and organization of word meanings according to their context of use), or differences in the nature and function of the parts of speech (the verb as the organizing principle, subject or object nouns as protagonists of the clause, adjectives as properties attached to a noun, etc.). Similarly, the notions of lexeme,[3] phrase, and sentence, which cover different linguistic realities, are not distinguished. The vector paradigm thus stumbles on the question of semantic composition, i.e., constructing a sentence's semantics on the basis of its structure, the semantics of the words it contains, and the nature and grammatical functions of those words. Indeed, given that the unit of modeling and representation is the word vector, a homogeneous and atomic entity, the organization of the different meanings, that is, the word's internal semantic structure, cannot be accessed by this paradigm. And any attempt to force the decomposition of the word vector, based on its coordinates in the whole space,

[1] Note nonetheless that certain linear models like HAL [15] assign each word a vector, which itself is made up of two parts, one representing the use of the word as a function of the word preceding it in the observation window, the second as a function of the word that follows.

[2] Note, on the other hand, that the aforementioned models of the *transformer* type, which generate context vectors for a given word, are currently being tested for their ability to organize its set of senses. But the conclusions across studies are contradictory and need to be reinforced.

[3] To refer to a lexical unit, I employ the term *lexeme* when the unit is a single word. At a more general level, I use the term *lexie* when the unit is either a single word (a lexeme) or composed of several words forming a "frozen" expression, as in the French locutions *train de vie* (lifestyle) and *savoir faire* (know-how).

by likening them to semantic features (see for example, [18]) was genuinely unable to come up with a satisfactory result. For the same reason, a model made up of only one representational entity cannot serve as a basis for representing the differences in nature and function among the parts of speech, since they themselves end up being neutralized, i.e., represented in the same format.

These inconsistencies are also found when we attempt to compare the vector paradigm grounded in an inventory of word uses drawn from texts, and the actual way humans implement language processes at the cognitive and neurophysiological levels. Several differences between real language processes and artificial neural models can be brought to bear [19]. Here, I will point out two. The first concerns the issue of the syntactic-semantic interface. While grammatical functions have been largely studied in linguistics, and although several neurolinguistic experiments (EMGs or EEGs) have indicated earlier processing of the distinction between the parts of speech than of the semantic content of each word in a sentence or in an expression, vector models, which are based on corpus scanning, did not initially separate syntactic features from semantic ones. The former are ignored in a learning process resting solely upon distributional regularities. Granted, some of the current research attempts to distinguish them *a posteriori* based on the specific distributional profiles induced by the parts of speech involved. However, distinguishing or classifying syntactic categories after the fact, with its substantial computational cost, does not account for their early neurophysiological processing. A second disparity concerns language acquisition. While a neural network learns solely from language uses recorded in large textual corpora, the child begins learning words by attributing content to them before acquiring them and mastering how they are used in the language. Moreover, the size of the corpora ingested by learning models and necessary for their performance largely exceeds the volume of language *stimuli* to which a child is exposed.

Lastly, the substantial number of parameters needed by these systems raises some questions. It seems legitimate to wonder whether this large number might not serve to compensate for the poor fit between the format of the units in these models and that of the linguistic processes. The sheer size of the means required to build them raises a critical question for society, namely the cost of the energy required, the reduction of which would be virtuous.

The fit between the models and lexical semantics, seen as a human science, is the very question that the present book addresses. My approach focuses on the study of content words.[4] The idea will not be to produce new mathematics but to juxtapose the idealities of mathematics and informatics with the representation of the entities and processes found in lexical semantics.

Toward an alternative proposal. Once I have determined the constraints impinging upon a model of lexical semantics capable of accounting for phenomena specific to this domain, I will propose some alternatives to the existing proposals. Three key points will be addressed. (1) The necessity, in modeling word meaning, of starting from another level than that of linguistic units. This level—of which I

[4] The study of function words, which are both frequent and few in number compared to lexical or content words, will only be addressed here indirectly.

describe the linguistic and neurolinguistic requirements and contingencies—will be called semantic matter. (2) The proposal that semantic matter is the result of establishing a linkage between systems of forms of expression (derived from writing or speech) and systems that are heterogeneous to those forms. (3) The proposal that semantic matter acquires its function in sentences by way of different states, the main examples of which will be studied here.

The present book is organized as follows. In the first part, I will present and support an argumentation that goes back to structural linguistics in order to show that the notion of sign and the duality between expression form and content form is echoed both in the search for a model in lexical semantics and in experimental in neurosemantic data. The distinction between paradigmatic axis and syntagmatic axis, another duality widely developed in this research trend, will also serve as a benchmark for questioning the relevance of current computational approaches grounded solely upon usage. The second part will be devoted to the link between the units of the language and the units of the model. The notions of semantic matter and the state of semantic matter will be presented. The idea of semantic matter/substance draws support both from a critique of the use of atomism in language modeling and from another critique regarding the use of the homogeneous, frozen approach in its totality (as in the use of vector spaces as a formal framework) in lexical semantics. The qualities and characteristics of this infralinguistic level will be presented in order to examine them in the light of suitable models. The third part will propose a new way of modeling in lexical semantics. It is rooted in the notion of a quasifractal tree-like structure. I will show how this framework satisfies all of the requirements previously brought to the fore and how it can be used not only to represent paradigmatic relations and syntagmatic agreement, but also their mutual interactions. I will stress the dual nature of semantic matter, which for each lexie necessarily contains an explicit part that is the form of the content (what is said), and in the shade, an implicit and inhibited part that refers to its opposite, contrary, or negation.

Semantic matter is modeled by a set of transformations corresponding to the nature and syntactic function of the lexie in the sentence. I will show how the formal structure, i.e., the tree-like structure, can act as "matter" to the basic states associated with their nature and function in the sentence. The case of the parts of speech—*verb, noun, adjective*, and *adverb*—will be presented, along with the differences between the subject and direct-object functions.

References

1. Geach, P., Black, M. (eds.): Translations from the Philosophical Writings of Gottlob Frege. Blackwell, Oxford (1952, 1960)
2. Russell, B.: On denoting. Mind **14**(56), 479–493 (1905)
3. Chomsky, N.: Studies on Semantics in Generative Grammar, vol. 107. Walter de Gruyter (2013)
4. Montague, R.: Universal grammar. Theoria **36**(3), 373–398 (1970)

5. Benzécri, J.-P.: L'analyse des données: l'analyse des correspondances. Bordas, Paris (1980)
6. Thom, R.: Stabilité structurelle et morphogénèse. InterEditions, Paris (1977)
7. Petitot-Cocorda, J.: Morphogenèse du sens. Presses Universitaires de France, Paris (1985)
8. Fellbaum, C. (ed.): Wordnet, An Electronic Lexical Database. MIT Press, Cambridge, Massachusetts (1998)
9. Landauer, T.K., Dumais, S.T.: A solution to plato's problem: the latent semantic analysis theory of the acquisition, induction, and representation of knowledge. Psychol. Rev. **104**, 211–240 (1997)
10. Ploux, S., Boussidan, A., Ji, H.: The semantic atlas: an interactive model of lexical representation. In: Proceedings of the Seventh Conference on International Language Resources and Evaluation (LREC'10), Valletta, Malta. LREC (2010)
11. Victorri, B., Fuchs, C.: Polysémie et construction dynamique du sens. Hermès, Paris (1996)
12. Bengio, Y., Senécal, J-Y.: Quick training of probabilistic neural nets by importance sampling. In: International Workshop on Artificial Intelligence and Statistics, pp. 17–24. PMLR (2003)
13. Mikolov, T., Sutskever, I., Chen, K., Corrado, G., Dean, J.: Distributed representations of words and phrases and their compositionality. In: Advances in Neural Information Processing Systems, pp. 3111–3119 (2013)
14. Devlin, J., Chang, M.-W., Lee, K., Toutanova, K.: Bert: Pre-training of Deep Bidirectional Transformers for Language Understanding. ArXiv Preprint ArXiv:1810.04805 (2018)
15. Burgess, C., Lund, K.: Modelling parsing constraints with high-dimensional context space. Lang. Cogn. Process. **12**, 177–210 (1997)
16. Mitchell, T., Shinkareva, S., Carlson, A., Chang, K.-M., Malave, V., Mason, R., Just, M.: Predicting human brain activity associated with the meanings of nouns. Science **320**, 1191–1195 (2008)
17. Huth, A, de Heer, W., Griffiths, T., Theunissen, F., Gallant, J.: Natural speech reveals the semantic maps that tile human cerebral cortex. Nature **532**(7600), 453–458 (2016)
18. Kintsch, W.: Predication. Cogn. Sci. **25**(2), 173–202 (2001)
19. Pulvermüller, F., Tomasello, R., Henningsen-Schomers, M.R., Wennekers, T.: Biological constraints on neural network models of cognitive function. Nat. Rev. Neurosci. **22**(8), 488–502 (2021)

Contents

Part I
Lexical Semantics, a Link Between Heterogeneous Levels of Organization

Chapter 1
Sign and Content Form: What We Have Learned From Structural Linguistics

Before structural linguistics, diachronic linguistics, which tracks the etymology of words from the joint standpoint of how they are formed and how their meaning evolves, served as research approach in lexical semantics. Bréal [1] proposed a definition of "the intellectual laws of language",[1] based on a delineated set of linguistic phenomena of which he demonstrated the effects and the explanatory power in the evolution and formation of meaning. Among other things, he included the specialization law, which results from our tendency to lower the cost of being understood by evading unnecessary aspects, and the distribution law ("la loi de répartition" in French), which is just as economical because it decreases lexical redundancies. The latter consists of redistributing the various senses across the synonyms in such a way that after being distributed the synonyms are no longer substitutable. In [2], Bréal tracks the competition between the French verbs *traire* (*to milk*) and *tirer* (to pull, to draw). *Tirer*, of Germanic origin, was substituted for *Traire*, which came before it and was employed earlier in a variety of contexts including *traire les cheveux* (to untangle hair) and *traire une aiguille* (to ply a needle). This drastically reduced the semantic scope of the earlier term.

Structural linguistics stood out from the approaches that came before it, which it accused of studying an accidental conglomerate "of facts" [3]. A founding element of this new analysis framework involved considering language as an organized whole. Saussure, the major protagonist of this movement, introduced the notion of *synchronic system* corresponding to a non-evolving state of a language whose structure could be studied. From this angle, language units are no longer viewed in terms of their language-specific diachronic trajectory but as a system of differences and mutual relationships. This system rests on the concept of sign, a composite entity made up of two heterogeneous and dual facets: the *signifier* and the *signified*. Accord-

[1] Note that this etymological approach, only skimmed over here, was itself, in its time, a source of controversies and innovation. Bréal [1] relied on the "human intelligence and will" at play in this diachronic process, there where his predecessors ignored it.

S. Ploux, *Semantic Matter and Its States*, Lecture Notes in Morphogenesis,
https://doi.org/10.1007/978-3-032-22809-3_1

ing to Saussure, the signifier is the auditory image of the word, and the signified refers to the concept or idea. Hjelmslev spoke of the form of the expression and the form of the content, while for Jakobson, it is the sensible (i.e., perceptible) and intelligible parts of a sign. I will come back to this later. Before continuing, let me present two introductory reasons why I chose the structural paradigm as the inheritance approach. The first has to do with the object of study, i.e., the semantics of lexie, the second to the objective of the modeling.

I could have contrasted this parent theory to other later linguistic theories, such as generative linguistics, the major research trend in the second half of the twentieth century. However, generative linguistics, as it was founded by Chomsky [4], eliminated the notion of semantic content to the benefit of an algebra of variables representing linguistic units in terms of their syntactic qualities. In this approach, then, the sentence *The cat catches the mouse* is analyzed as the result of a series of rules (or by a tree equivalent to their implementation) that would be identical (save for the attribution of the final values, i.e., the vocabulary words) to that of all other sentences with the same syntactic format (e.g., *The sergeant eats an apple*; *The moon hides the sun*). The emphasis here is on the organization of the sentence being governed, and it marginalizes the role of semantics, particularly lexical semantics. If it then became possible to include semantic aspects in this paradigm by Chomsky [5], the inclusion rests on the idea that the interpretation of a sentence is derived from the application of rules. The grammatical structures defined in the deep structure determine that interpretation. Other authors pursued the generative approach in semantics without a logical framework. In this trend and in line with Frege [6], Montague [7] proposed a model for interpreting the semantics of a sentence. The model in question is the first-order logic. A truth value is assigned to each proposition, and referents (the world entities pointed to by the words) are assigned to content words. At the present time, this approach remains marginal, and it eliminates—just like generative grammar with which it shares the axiomatic and deductive framework—the phenomena of lexical semantics and semantic relations between lexemes, to the benefit of reference. Moreover, the hypothesized autonomy of language when compared to the totality of cognition [8], presupposed by Chomsky, ignores *de facto* the link between a form of expression and a content.[2] So this tradition does not take the lexicon, its structure, or the differentiated study of word semantics into account. Because the object of study in the present book is lexical semantics, the structural approach involving the link between a lexie and its content thus constitutes a scientifically privileged inheritance. Despite this constituent choice, I will nevertheless come back to the importance of the different syntactic natures and roles and their lexical anchoring, which have rarely been taken into account or have even been evaded in the structural movement.

[2] Note that Chomsky's hypothesized autonomy of language, which is a hypothesis of modularity and encapsulation, differs from Saussure's hypothesis, wherein a language is disconnected from its psychological anchoring. The former separates language from all other cognitive processes; the latter, via the notion of sign, situates languages between expression and shared ideas.

Sign and model. Lastly, the claimed inheritance of structural linguistics shows up again in the modeling process. Indeed, modeling lexical semantics requires assigning, to a given form of expression, a mathematical or computational entity that will represent its meaning. In current computational approaches, this formal entity can be a vector or a spatial form in geometric models. These various embeddings in a shared space makes it possible to establish relationships between the meanings of words by calculating how close or how far they are from each other and is thus a system of relationships between units in the same way as the signs of structural linguistics. For the two reasons mentioned above—what domain is being studied and how it is modeled—I will look further into the notions of sign and content form, in order to draw support for and further specify what is at stake and the nature of a suitable model.

1.1 Sign and Content, a Taciturn Affinity

As recalled above, two facets (which in line with Hjelmslev, I will call the expression form and the content form) and how they are situated with respect to each other in the sign constitute, from a structuralist perspective, the underlying principle of language. Saussure stressed the arbitrary, conventional nature of the link that relates subdivisions issued from heterogeneous substrates. If this link is arbitrary, the construction of the sign is explained as the effect of mutual operations and adjustments bearing on the spaces in which expression and content are rooted.

Facets of the sign: sensorimotor materiality and immateriality of content. The unity of the sign is controlled in this theory by the operations of substitution, and contextualization, and their effect on the disparity *vs* equivalency of the expression and content forms. Accordingly, the different ways of pronouncing a word by varying the speech rate, intonation, or quality of the sounds will often have no effect on the word's identity. In a dual manner, a change in content can be expressed by the same lexical form. For example, in the French expressions "la force du vent" (the force of the wind) and "à bout de force" (out of force), the same concept (force) coincides with the same phonological slice /fors/. Saussure concluded that in such a case, we are dealing with one and the same linguistic unit. In contrast, in "il me force à parler" (he forces me to speak), the word "force" has a totally different meaning. So even though they are phonologically identical, these two segments are different linguistic units [9].

The expression form is anchored in a sensorimotor substrate whose materiality can be relatively well accessed by studying the sounds, the articulatory movements, and the form of the buccal cavity. The expression form does not conform directly to this substrate but is derived from regularities operating upon it. In other words, an expression form is a unit in a sensorimotor space "quotiented" by the classes of regularities specific to the language.

In contrast, the content form is tied to a more abstract substrate, of which the early structuralists gave less complete descriptions. This flaw may partly result from differences across authors in the notion of content. For Saussure, the content of a sign is seen as a concept, an idea belonging to a collective consensus, insofar as a language is not an object to be understood from an individual point of view, but from a social point of view. For Jakobson [10], the signified refers to the part that is intelligible and is thus fundamentally anchored in mental processes, and even in cerebral processing rooted in individual functioning. Note that a large body of recent studies in the neurosciences aligns with Jakobson's point of view. They have uncovered some neurophysiological correlates triggered by the reading or hearing of words. The cerebral activities involved in the processing of these lexical tasks take place in the language areas of the brain, which are located around the Sylvian fissure and also in the areas that process a written or heard-word's content. For example, the reading of a color noun (e.g., "red") will trigger activities in the occipital brain areas involved in vision, while that of an action verb ("hit" or "pick") will trigger processing in the motor and premotor areas involved in performing the corresponding movement [11]. The link between the auditory or visual processing (which can be likened to the processing of the word's expression form) and the brain activities induced by that processing and associated to its content could therefore be a manifestation of the content form *à la Jakobson*. Furthermore, as with the expression form, the content form can only be defined by regularities operating on neural activities, in such a way that all of the neural dynamics resulting from the perception of a color must be *quotiented* by a delineated group of variations in order to enable the content form specific to the word "red" or "green", for example, to be assigned.

In sum, for the early structuralists, the lack of a consensus on the essence of the content substrate (whether neural or social), the shared sense that its nature is amorphous, and difficulty grasping it in a systematic or formal way led to a desire to remove it from the study of "language": "[Linguistics] must not attempt to grasp language as a conglomerate of nonlinguistic events (physical, physiological, psychological, logical, sociological), but as a self-sufficient whole, a sui generis structure" ([3], our translation). Language, then, becomes a system of signs treated as units with two non-decomposable facets equipped with relationships that link them to each other and organize the system.

The next section will deal more specifically with the notion of content form, i.e., the signifying facet that ties the expression form and its content to each other. I will also look at the way in which—in line with Jakobson's and Hjelmslev's work—the content form became describable and formalized by relying on a metalanguage.

1.2 Content Form: Theorizing and Modeling

In order to introduce how Saussure grasped the content form in his Cours de Linguistique Générale[3] (Course in General Linguistics) [12], I have schematized two figures (see Fig. 1.1). The first (Part 1 of the figure) has the structure of a coin whose two indissociable sides represent the signifier and the signified; the second (Part 2 of the figure) shows sections cut out jointly from the two substrates, as with a cookie cutter. However, the features of arbitrariness and unity—which are pillars of the notion of sign for Saussure and ensue from the fact that its content form has no subparts that would link up to a subpart of its expression form and vice versa—led me to modify the figure by proposing to schematize the sign as a double cone each one standing for both of the sign's facets, and their embedding in heterogeneous substrate spaces. Their point junction at their summits illustrates the arbitrary and unitary features. For Saussure, the content form—which relates the expression form to content (corresponding to the upper cone in Fig. 1.1, Part 3)—its extent, and its own structure are neither theorized nor modeled. As recalled above, there is no consensus that could be called upon to explain this lack. However, its implications for the study of lexical semantics are real and impede the structuralist conception itself. If the reasons for this follow—as I believe—from the vagueness surrounding the nature of content, any attempt to design a semantic theory necessarily has to rely on the nature of this link. Indeed, the relationships of difference between units or between the paradigms that group them together depend upon this link, fundamental for structuralists. Accordingly, differences between signs that mark the distinctions between contents should find a correspondence expressed as a difference between what are represented here as content-form cones and the anchoring of their bases in their respective contents (Fig. 1.2a). Likewise, a paradigm (sharing of content) (Fig. 1.2b) should correspond to the sharing of respective content forms, and a phenomenon of polysemy could correspond to an extended structure of the content form (see Fig. 1.2c). The idea of using a conical form, which serves here to stress the need for a model that adheres to the arbitrariness and unity properties of the sign, is a sketch, a stage, of the model that will be developed in the upcoming chapters.

To provide a remedy for this deficiency while still hypothesizing autonomy of the object "language" with respect to the content substrate, Jakobson and Hjelmslev introduced the notion of metalanguage. The idea was to formalize the relations between units by working from a level other than that of the language. Greimas [13], by freeing himself of the concept of sign, Rastier [14], and many linguists pursued this endeavor using the notion of seme or semantic feature. The feature, atomic, must be something other than a unit of the language and allows for an analysis of the entire lexicon and all texts. For certain authors, it must be universal. Their proposal of semes results from the evaluation of the semantic differences between words taken two at a time. These atomic elements are assumed to be finite and few in number.

[3] "A linguistic event can therefore be pictured in its entirety—i.e., language—as a series of contiguous subdivisions marked off on both the indefinite level of jumbled ideas (A) and the equally vague level of sounds (B)" ([12], page 181, our translation).

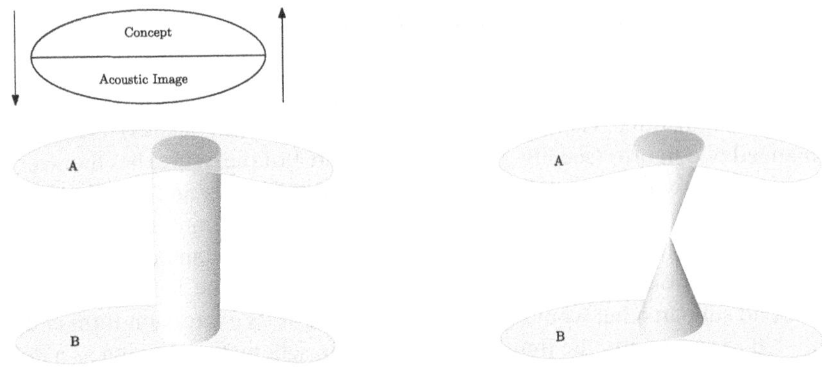

Fig. 1.1 Upper (1): depiction of the sign. Lower left (2): breakdown of the sign at the idea level (**a**) and sound level (**b**). Lower right (3): depiction of the arbitrariness of the sign

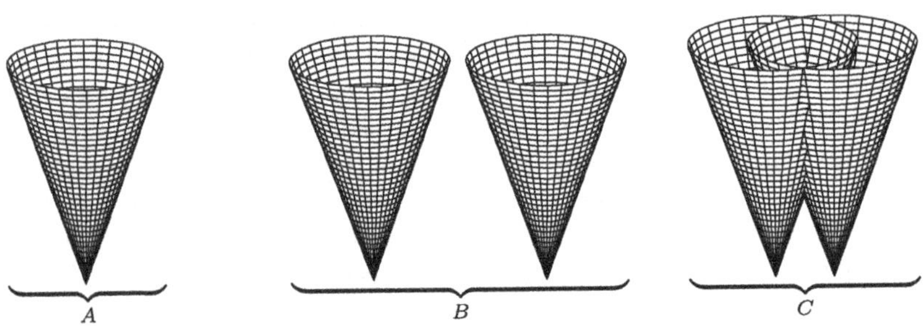

A. Polysemy depicted by the scope of the content form associated with the sign
B. Two disjoint content forms
C. Two content forms with partially shared content

Fig. 1.2 Different configurations of the relations between content forms

An often quoted example taken up by Pottier [15] is the breakdown into features of words about "seats", i.e., things one sits on (see Fig. 1.3). The formulation often retained uses arithmetic symbols. One adds and/or subtracts features in tables in order to "calculate" the lexical semantics and textual isotopes (bundles of consistent features that bind and give cohesion to a text).

To this day, we still do not have a list of features that has made it possible to implement a systematic, comprehensive calculation of the semantics of words or isotopic phenomena. This unfinished nature[4] raises several questions. The first concerns the methodology used to construct the required finite lists. All proposals so far are based on examples obtained through an expert evaluation of the shared or differentiating

[4] Note that for the purposes of modeling, being unfinished does not preclude the use of this decomposition method in the semantic analysis of texts (literary, etc.).

armchair = *"with arms"* + *"with a back"* + *"for sitting"* + *"a person"*

chair = *"without arms"* + *"with a back"* + *"for sitting"* + *"a person"*

sofa = *"with arms"* + *"with a back"* + *"for sitting"* + *"several persons"*

bench = *"without arms"* + *"without a back"* + *"for sitting"* + *"several persons"*

Fig. 1.3 Breakdown into semes of words referring to seats (taken from Pottier [15])

features. But beyond the obvious difficulty of formalizing this methodology, the question that arises is whether it is even possible to do so. Cleary, there are two formal prerequisites: (a) the need to validate the completeness of the list of features (i.e., the existence of a finite list capable of covering the entire set of relations between words) and (b) its consistency, i.e., the intrinsic lack of contradiction between the set of features on the list. To my knowledge, these necessary prerequisites have not really been considered. Moreover, the idea of calculating features also raises the question of the homogeneity of the feature space. Are all of the features of the same nature? And if not, how can we understand that they can be added, or that their sums enable the calculation of the semantics of words which themselves are of different natures, such as adjectives, nouns, or verbs? The fact that these questions—which ensue directly from the choice of an arithmetic calculation paradigm and the characteristics specific to that choice—have not received exhaustive answers and therefore raise some doubts about the relevance of this approach for developing a formal model.

In my mind, however, beyond these objections, one of the essential interrogations regarding this approach resides in the idea that features, as metalinguistic units, may not be units of the language but are rooted in content or content forms. For the mere fact of stating so does not suffice to mask the artificiality of this postulate, given that features are always expressed in words.

This chapter was aimed at highlighting the vagueness of the notion of content, as well as the difficulties that surround it and the formalization of the content form. It will allow me, in the next chapter, to address the question of the relations between and among the units of a language. In the structural approach, the relation between the signifier and the signified, for a given sign, hinges on the other signs and thus on the type of link between their content forms.

References

1. Bréal, M.: Les lois intellectuelles du langage fragment de sémantique. Annuaire de l'association pour l'encouragement des études grecques en France **17**, 132–142 (1883)
2. Bréal, M.: Essai de sémantique, science des significations. Hachette (1904)
3. Hjelmslev, L.: Prolégomènes à une théorie du langage. Arguments **35**, (1971)

4. Chomsky, N.: Aspects of the Theory of Syntax, vol. 11. MIT Press (2014)
5. Chomsky, N.: Questions de sémantique. Éditions du Seuil (1975)
6. Frege, G.: Sens et dénotation. Écrits logiques et philosophiques, pp. 102–126 (1971)
7. Dowty, D., Wall, R., Peters, S.: Introduction to Montague Semantics. D. Reidel Publishing Company, Dordrecht (1981)
8. Piattelli-Palmarini, M., Piaget, J., Chomsky, N., Noizet, Y.: Théories du langage, théories de l'apprentissage. Points (2020)
9. de Saussure, F.: Cours de linguistique générale. Bibliothèque scientifique, Payot, Paris (1972)
10. Roman, J.: Essais de linguistique générale. Minuit, Paris (1963)
11. Jeannerod, M.: Language, perception and action. How words are grounded in the brain. Eur. Rev. **16**(4), 389–398 (2008)
12. de Saussure, F.: Course in General Linguistics. Philosophical Library, New York (1959)
13. Greimas, A.J.: Sémantique structurale: recherche de méthode. Presses universitaires de France (2015)
14. Rastier, F.: Sémantique interprétative. PUF, Paris (1987)
15. Pottier, B.: Sémantique générale. PUF, Paris (1992)

Chapter 2
Paradigmatic Relations, Syntagmatic Agreement, and Content Form

This chapter deals with paradigmatic and syntagmatic links. It addresses how the definition of the sign can have an impact on them, and what they say in return about the content form. These two types of links remain at the core of current research, particularly in approaches grounded in automatic text-based learning and thus, on syntagmatic agreement. I will begin by attempting to show how these links have been approached and how more attention has been paid to one or the other in different domains including philosophy, linguistics and computational linguistics, psycholinguistics, and more recently, neurolinguistics. I would like to demonstrate why and how the duality they form is a necessary one. This second demonstration will provide a critical argument about many computational approaches. Next, I will come back to the notions of unit and minimal unit, which will enable me to show that I depart from the idea that language units are at the level of, and have the necessary granularity for, formalization in lexical semantics. The development of these two anchoring and differentiating points (i.e., differentiating with respect to structural linguistics) will allow me to establish the proposed model in the remaining chapters of this book.

In addition to the notion of sign, Saussure introduced two types of relations between lexical units. The first is the relation that links up words located in the same phrase (*in praesentia*), which he called syntagmatic. The second is the relation that links up or opposes a linguistic unit to other units that have features in common (*in absentia*). For example, irrespective of the fact that the following words are not part of the sentence *The cat catches the mouse*, the words *carnivorous*, *hunt*, and *rat* (or their contents) can each be associated, in a subject's memory, to each word in the sentence. Saussure called this type of relation associative, a term which today can lead to confusion because it was used later to qualify syntagmatic association links or regular co-occurrence links (such as *car-speed*, *mountain-snow*). Saussure described the associative relation as nebulas of words compounded on the basis of semantic proximities and similarities of form. In this view, the paradigm can be based on the word's root (e.g., for the word *teaching*: *teach*, *teaches*), its morphology (*aligning*, *changing*, etc.), its meaning (*instruction*, *education*, etc.), or founded solely on audi-

S. Ploux, *Semantic Matter and Its States*, Lecture Notes in Morphogenesis,
https://doi.org/10.1007/978-3-032-22809-3_2

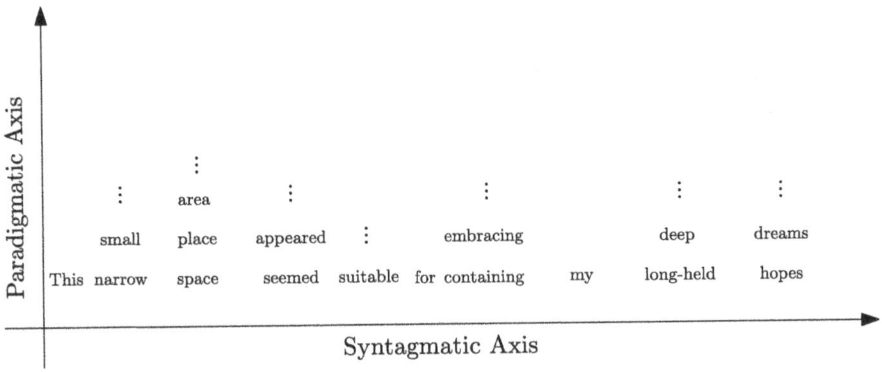

Fig. 2.1 Depiction of the paradigmatic axis (vertical) and the syntagmatic axis (horizontal). This example was derived from a quotation found in Chateaubriand's Mémoires d'outre-tombe ([1], p. 5)

tory similarities (*teaching*, *breaching*). The duality between the two types of lexical relations was taken up and re-delineated—and their definitions systematized—by later structuralists. For Hjelmslev, the term "paradigmatic" could be renamed "system" (or language) and the term "syntagmatic" could refer to production processes (or text). Jakobson (who remamed them "axes") assigned them unilaterally to one or the other axis, according to their governing process. He called the syntagmatic axis the combination axis, and the paradigmatic axis the selection axis. It is in the latter approach that we find these links depicted, in the standard way, as vertical and horizontal axes (Fig. 2.1). Their definition, which currently remains fixed in the literature (and is the one I use here) goes as follows: Two units u and u' belong to a same paradigm if and only if they can be substituted for each other in one and the same syntagmatic schema, that is to say, if there exist two phrases vuw and $vu'w$ that are similar and valid in terms of a given criterion. Assessing the similarity between the two phrases is left up to the expertise of the linguist. This assessment relies *in fine* on the linguist's knowledge of the language being described and his or her access to the cognitive contents linked to the phrases and the words they contain. Hereafter, I will use the term *relation* to talk about paradigmatic links and the term *agreement* to talk about syntagmatic links. This terminological distinction highlights the difference in the nature of the links underlined by Hjelmslev and Jakobson.

Although syntagmatic agreement can be accessed directly via a set of valid productions in a given language, the paradigmatic link requires a theory of the lexicon and its structure. It is presented in terms of the operations of substitution and contextualization, and their stabilizing effects, or, on the contrary, their altering effects. These are precisely the operations mentioned in the preceding chapter to describe the sign as a minimal unit. Although the structuralists made grammar into a derivation of these operations rather than a primary organization, the fixed stability criterion requiring content-form conservation can be generalized to conservation of the syntactic form. Thus, if the idea is to preserve the syntactic schema and its conformity, then the adjectives *profound* and *unfortunate* will belong to the same paradigm, since "Of some profound word every man is the disciple"[1] and "Of some unfortunate word every man is the disciple" are two valid and syntactically equivalent phrases. Concerning the academic research on semantic paradigms, the examples commonly mentioned cover links of similarity (synonymy or hyperonymy) or on the contrary, opposition (antonymy), with the grammatical category being preserved. For instance, the conservation criterion can be satisfied by a substitution that leaves the meaning of the phrase unchanged (or hardly modified). In this case, the two phrases "a bright day", "a luminous day", or "a clear day" which are paraphrases of each other, are variants of one and the same schema in which *bright*, *luminous*, and *clear* are substitutable and thus linked by a paradigmatic relation. This last criterion even constitutes one of the defining criteria of synonymy.

Lastly, note that while these links were initially defined based on substitution or combination operations and their linguistic effects, their relevance in accounting for the mental processes underlying language was also stressed by Jakobson. He showed [2] not only that their duality was at play in the cognitive and neurophysiological processing of the lexicon, but also how a deficiency among patients in one of these links is tied to one of two known types of aphasia. A person with a paradigmatic deficiency has trouble selecting words, while a person with a syntagmatic deficiency has difficulty combining words to form sentences. Other more recent studies [3] using brain imaging support the hypothesis that the organization of the mental lexicon is based on the complementarity of the paradigmatic and syntagmatic axes.

Two questions will be addressed in the rest of this chapter. The first deals with the anchoring of these two types of links in the content form of linked units. The second deals with their complementarity. The idea will be to show that their duality is necessary and that neither type of relation can be substituted for the other. The latter demonstration provides a critical argument against current computational approaches based solely on co-occurrence links or co-occurrence patterns and thereby ignore paradigmatic links, as well as approaches that rely on lexical ontologies and ignore usage and thus, syntagmatic links.

2.1 Lexical Links, Sign, and Content Form

Paradigmatic relations, such as synonymy, are determined as a last resort on the basis of an assessment of a more or less partial overlapping of content. Indeed, assessing similarity, opposition, or equivalence in a given context requires expertise grounded in knowledge of the language related to the contents in question. As recalled in the preceding chapter, it seems surprising when we read texts by the early structuralists that the dual nature of the sign as being made up of an expression form and a content form did not serve as a basis for studying the organizing axes of languages—

[1] Our translation of Victor Hugo's "De quelque mot profond tout homme est le disciple.", Les contemplations.

understood to be the basis of relationships between signs or classes of signs—without recourse to the interaction between the content forms that those relationships involve.

How can we understand this obstacle, which prevents taking advantage of the sign's duality? Two reasons seem to explain or even clarify the hiatus between "sign" and "lexical relations". The first follows from the fact that, for the early structuralists, the content form, as a facet of the sign, had no substance of its own. It therefore could not be the material for calculating overlaps or differences between language units. One could object that Hjelmslev's search for a metalanguage, pursued by other authors, was aimed precisely at finding a way to systematically fill the gap between expression form and content. This approach, which is not really complete from a formal standpoint, as recalled above, rests on the performative presupposition that the features always verbalized and defined in a language are not language units. The result of this manifest contradiction is that they cannot be content forms, for a content form is not a sign (a language unit) but one of the two facets of a sign.

The other reason rests on the accepted "given" that "language can only be explained by language" ([4], p. 135). A content form linking an expression form to nonlinguistic content, would—by the very nature of the linked content—contain some trace of a nonlinguistic organization, which would make it unusable in a scientific approach based on the interplay of differences and links between signs. While avoiding this presupposition, the paragraphs that follow attempt to promote a possible understanding of the content-form interactions induced by paradigmatic and syntagmatic links.

Paradigmatic relation and content form. Regardless of what nature one grants to content and how one approaches paradigmatic relations, the latter necessarily rely on either the sharing of content forms or, in the case of opposition or differences, traces of these distinctions in the organized system of content forms. Accordingly, determining what remains the same in a substitution of two signs in a phrase or sentence involves selecting what their respective content forms have in common and disregarding (or in the case of opposition, looking at) what distinguishes them. For example, the substitution of *bright* for *intelligent* in the phrase "a bright student" (*vs* "an intelligent student") will select a part of the content form of *bright* that is shared by the content form of *intelligent* while ignoring their unshared or unmatched meanings in the phrase. This part is different from that selected in the substitution of *bright* for *luminous* in the phrase "a bright day". Now in the case of an opposition (for example, *black vs white*, *kind vs mean*, *enter vs exit*), the differences can only be understood within a set of content forms anchored in contents related, respectively, to colors, personal qualities, and movements. The antonymy relation reveals the need for an underlying organization in which a shared referent must first be found and then an opposition within that referent must be made. Note que these selection and substitution operations do not necessarily involve symmetric relations between the content forms of the lexical units. For instance, take the sentence "What could be more dismal than a party in dark rooms lit by the turquoise sky at the end of day, in spring?" (Blanche, Mes Modèles, 1928, p. 133, our translation). The hyponym *turquoise* (used here as an adjective) can be replaced by *blue* (a hypernym), which is less precise but not conflictual, whereas the reverse could lead to an imprecision because saying "the blue river" does not indicate that the river is turquoise. These two phenomena (opposition and asymmetry) highlight the fact that paradigmatic relations rely on an organization of a system of content forms that is at least partially hierarchized.

The partial overlapping of the content forms of lexies, and the plurality of possible ways of sharing for a given lexie echo two dual phenomena that are crucial in lexical semantics: polysemy and semantic stability. Polysemy refers to the semantic multiplicity of a lexie and the selection made among this diversity in a given syntagmatic context. Polysemy thus involves an operation of matching by selection between the content form of the lexie and the bundle of phrases that contain it. So the word *bright*, as recalled in the preceding paragraph, has different meanings, and the context (lexical and syntactic) in which it will be employed in this diversity will determine the correct meaning or meanings. More generally, this phenomenon applies to all units in a given phrase or sentence for each one of which a part of the content form will be selected and the rest will be ignored. Pustejovsky [5] goes even further by proposing that not only selection can occur, but also creation of meaning, or in our approach, extension of the content form. It thus seems that it is not just the sign itself but a part of its meaning, that is, a part of its content form, that is selected or even created and takes effect in the calculation of the meaning of the phrase or sentence that contains it. In a dual manner, the semantic stability of a lexeme—in addition to the multiple uses and different meanings brought to bear—would be based on their incorporation into one and the same canvas, the organizing core of the content form whose variations in context would amount to adjustments or unfoldings. This view can be likened to Culioli's [6] notion of schematic form, which for each lexical item, interrelates invariance and the internal and external variation modalities.

We have arrived at the conclusion that the content form cannot be composed of linguistic units. This leads to the hypothesis that an infra-linguistic level is necessary to account for paradigmatic relations. This infra-linguistic property is derived from the fact that the content form's constituents have a finer granularity than the sign, since the sharing that takes place between content forms points out partial overlapping that is finer than the content forms of linguistic units. For each of these two phenomena (polysemy and semantic stability), the selection process that applies to the selection of a lexeme's meaning in a given co-text therefore cannot depend solely on lexical units, even if they are taken as units of a system, but must be seen as links between their content forms, which themselves are embedded in a system that encompasses them. If the paradigmatic relation is intrinsically imprinted in the space of content forms, what about syntagmatic agreement?

Syntagmatic Agreement and Content Forms

Phrases, sentences, and arbitrariness. Unlike the sign, which is not the locus of a morphism between expression form and content form, the structure of the sentence (or phrase) is partially homologous to the structure of its content and thus of

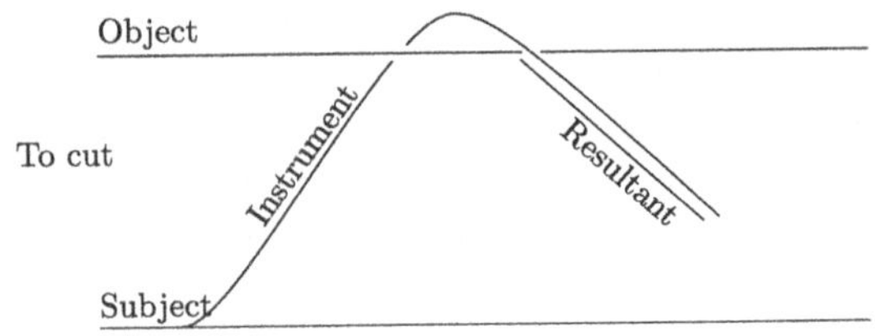

Fig. 2.2 Thom's bifurcation diagram of the verb *to cut*, with its actants, the subject, the object, and the instrument, and the resultant of the cut, [8], p. 213

its content form. Accordingly, Tesnière [7] viewed the organization of sentences as taking place within a verbal scenario. The actors on stage are the subject and the complements of the verb; the process expressed by the verb defines their interactions; the circumstances of the scene (time, place, and manner) are expressed by the circumstantial complements. In contrast to the notion of verbal scenario as a model of the sentence that remains both highly generic and metaphorical, Thom ([8], pp. 188–189) proposed a dynamic model describing how sentences are organized around specific verbs. Verbs are schematized in terms of a dynamic, verb-specific interaction between its arguments, in line with the nonverbal interaction at stake. For example, the verb *to cut* is modeled as a spatiotemporal form derived from a parabolic umbilic that depicts the scene and the interaction specific to this verb and its arguments (the subject, the object that gets separated, the instrument) (Fig. 2.2).

As an initial assessment (which I will come back to later), one could consider such dynamic schemas as a way of modeling the verb's content form in terms of an organizing node of the sentence, since such schemas make the connection between an expression form and a multiply-manifested content by extracting the essential regularities and ignoring any contingent elements. Contrary to the arbitrariness of the sign, these different proposals thus stress the presence of morphisms between the level of the expression and that of its content whenever a sentence or a phrase is at stake. However, they fail to consider the variability of the link between each unit of the sentence and its content. So depending on its usage context, the verb *cut* can refer to a division (*cut the bread*) as depicted in Fig. 2.2 or an interruption (*cut the speaker short*). The study of semantic composition deals with this aspect.

Syntagmatic agreement and composition. Semantic composition is traditionally defined as involving the calculation of the meaning of a phrase or sentence by looking at combinations of the meanings of its constituents and the syntactic structure that links them. As mentioned above, syntagmatic agreement (or usage context) perform selection or extension operations on the content forms of a sentence's constituents. While these transformations are anchored in the constituents' content forms (i.e., the semantic matter), they are also constructed according to their respective roles (process, actants, circonstants, etc.) within the sentence. For instance, in the sentences "He finished a substantial amount of work", "This substantial amount of work finished him", switching the position of the pronoun and the subject or object of the word work changes the meaning of the verb *to finish* and the type of interaction it underlies.

The point of view developed here is that determining the semantic composition of a sentence consists of modeling its content form as a function of its expression form in the utterance language, of the content form of its constituents, and of the content form associated with their respective functions in the sentence. As noted by many authors, other factors also have an impact on content forms, which, for some, are conventional and for others are rooted in cognitive processes. As Saussure reminded us, in the repetitive utterance "Gentlemen! Gentlemen!" the two instances of Gentlemen are not absolutely identical semantically. Repetition can also have an expressive effect, frequent in Mandarin, and is present in childish naming in the form of syllable repetitions (*mama*). Another example: screaming the word "Come" rather that simply saying it, or using different fonts as done in comic strips, will have an effect on the meaning of the verb, and can be paraphrased as "Come and hurry up". Similarly, word order and rhythm can also represent temporality, as in the quote "veni vidi vici". The present book will be limited to the lexicon and the roles of the units in a sentence. I will not address the effects of prosody, repetition, and so on. Note that contrary to what was said above regarding semantic composition, the freezing of a phrase (e.g., *show to the door*) or a syntactic group like "pomme of terre" in French (literally, apple of the earth) is an operation that weakens the morphism between the expression form and the content form by making the link more compact and more arbitrary. In this case, the expression form is still a composite, but the content form no longer adheres to this compounding, or does so in a less substantial way.

Imprinting of syntagmatic agreement in the content form. At the beginning of this chapter, I recalled that a content form contains the trace of the paradigmatic sharing of content in the case of lexies that are similar in essence (e.g., *bird-canary*, *bright-luminous*), and that syntagmatic agreement is expressed in the effects of selection or extension operating on the content forms. In line with the bond they weave between the expression level, the content forms necessarily contain a trace of the type of interaction existing between units in these two levels. For instance, the content form of the lexeme *freedom*, will contain the imprint of possible ways in which the content associated with freedom interacts with other contents (say, those expressed verbally such as *take away*, *give back*). Another example is the content form of the word *child*, which contains the imprint of all interactions carried out by the child (*play*, *eat*, *laugh*, *sleep*, *cry*).

In sum, I set out in this chapter to show that paradigmatic and syntagmatic links are imprinted in the content forms of a language's constituents (lexemes and lexies, phrases, sentences).

2.2 Necessary Complementarity of Syntagmatic and Paradigmatic Links in Lexical Semantics

Paradigmatic and syntagmatic links are also found in two major research trends in lexical semantics: the ontological or Aristotelian[2] approach and the usage-based approach. In the former falls in the lineage of, and claims to be derived from Aristotle, attempts to represent the semantics of a word in terms of a concept: the content of an abstract thought that is separate from the reality of a thing, situation, or phenomenon. In the latter approach, word usage in the language is a possible route for accessing meaning. By putting in perspective the ontological approach and paradigmatic relations on one hand and the usage approach and syntagmatic agreement on the other, we can attempt to learn from the dispute between them so as to stress their necessary complementarity.

2.2.1 Ontological Approach and Paradigmatic Relations

The ontological approach is based on the notion of concept, and in linguistics, that of lexical concept. The idea is to address the meaning of a word by characterizing the concept to which the word refers. To do so, the concept is defined in intension by the list of attributes or features that define the necessary and sufficient conditions an item must satisfy to belong to the category delineated by the concept, e.g., *has a beak, feathers, wings, etc.* for the concept *bird*. In psycholinguists, this paradigm led to the proposal that concepts are organized according to a hierarchy insofar as a sub-concept (or hyponym) such as *canary* inherits of all the properties of a supra-concept (or hypernym). Each concept or node of the ontological hierarchy has characteristics of its own, like the fact of having yellow feathers for the concept *canary*. This model is inferential in that the necessary and sufficient conditions for belonging to a concept can be determined and inherited. This implies that because the properties of *canary*, for example, are accessible and inherited, and can be inferred from the properties of the subsuming concepts ($canary \sqsubset bird \sqsubset animal$). The hierarchical tree structure that traces inheritance links has been validated experimentally. Collins and Quillian [9] showed that the time needed to process sentences of the type "a canary is yellow", "a canary can fly", "a canary breathes" increases and is positively correlated with the length of the path in the hierarchy between the concept (*canary*) and the supra-concept that has the inherited property (here, *is yellow* for *canary*, *can fly* for *bird*, *breathes* for *animal*). This hierarchical organization is also consistent with clinical observations showing that certain brain lesions induce impairments (in tasks such as naming or pointing) that selectively affect a broad category such as the set of living *vs* non-living entities, or finer categories like parts of the body and fruits and vegetables (see for example [10]).

---

[2] "There is a science which studies Being qua Being, and the properties inherent in it in virtue of its own nature." ([24], Book 4).

This theoretical framework has also been highly criticized on the basis of other experiments demonstrating a prototypical organization centered on items representative of a category (e.g., *robin* for *bird*, *dog* for *animal*). For one, Rips et al. [11] showed that not all exemplars of a concept are equivalent as they would be if a single list of necessary and sufficient conditions existed. A concept's internal organization is more like a membership gradient constructed from shared attributes, like a family resemblance [12]. In this approach, a prototype has a large number of properties in common with other members and a small number of properties in common with members of another category. For instance, a *robin* (prototype of the concept *bird*) *has feathers*, *has wings*, *has a beak*, *can fly*, *lays eggs*; a *penguin* (a non-prototypical exemplar) *does not fly*, *knows how to swim*, and so on. However, outside of this controversy, the ontological approach has made a substantial contribution to the study of lexical semantics. This kind of inheritance already existed for Aristotle, who defined synonymy as identity of notion [13]. The famous WordNet database [14] via which Miller, its author, planned to create a dictionary based on principles derived from Aristotelian psycholinguistics, relies on the notion of a Synset (set of synonyms) representing a lexical concept. WordNet is organized into hierarchies based on hyperonymy or hyponymy links.[3] The WordNet project, however, never met the requirements of the Aristotelian tradition of associating a list of necessary and sufficient attributes to every lexical concept, that is, a logical and inferential definition. On the other hand, a lexical concept is constructed as a paradigm, in the sense used in structural linguistics, since it is represented by a list of synonyms (Synset) that are equivalent by substitution in a given set of usage contexts.

2.2.2 Usage-Based Approach and Syntagmatic Agreement

Against the grain of logical determination, Wittgenstein stressed the fruitless, impossible, and unnecessary attempt to characterize the meaning of words via a formal definition.[4] He states, "The meaning of a word is its use in the language". This criticism of the ontological paradigm was accompanied by an alternative proposal involving the notion of family resemblance, a sort of binder between a concept's exemplars deemed to replace the notion of definition. Let me stress here that the notion of family resemblance is grounded in a genomic metaphor which, if fully panned out, would involve a primitive level heterogeneous to language but enabling its realization. This level, then, would be substituted for a list of manifested, necessary and sufficient properties. This same criticism showed up in psycholinguistics with the work initiated by Rosch and others cited above, and also in linguistics. Many linguists in the late 1980s and early 1990s defended the idea that the notions

[3] Other links were added later to include lexical relations that escaped this organization, such a meronymy, antonymy, and troponymy for verbs.

[4] "When I give the description: 'The ground was quite covered with plants' do you want to say that I don't know what I am talking about until I can give a definition of a plant?" [21].

of prototype and family resemblance developed in the study of categories applied to lexical polysemy (see for example [15]).

This trend—usage-based semantics—also spread to computational linguistics. LSA, HAL, and other models [16, 17] circulating in the years 1990–2010 and a lot of research, especially current artificial neural networks, is based on the same arguments. These systems perform a scan of large corpora (sets of uses) to build models of lexical and text semantics. Usage and syntagmatic agreement correspond to exactly the same thing. It should be noted that the syntagmatic agreement in which the above models are rooted is a regular, co-occurrence link that ignores both syntax and the organization of the lexicon, which these models attempt *a posteriori* to find imprinted in computed vectors. As far as semantic similarity is concerned, an argument put forward to support this approach is that it arises from a community of uses. Let me give an example illustrating this argument. In speech or in written texts, one can find phrases such as "many *resemblances* between the testimonies", or "many *similarities* between the testimonies". Consequently, text scanning by linear methods (data analysis) or non-linear methods (neural networks) that detect co-text regularities enable the representation of similarities (here, between *resemblance* and *similarity*). The vector-based representations obtained do indeed provide evidence of semantic proximity phenomena. But in a homogeneous vector space, the list of a word's neighbors, i.e., words whose vectors are the closest to each other, make no distinction between strong semantic similarities (synonymy and more generally paradigmatic links) and strong associations (recurring syntagmatic links such as *perfect-match*). And also, while a computed, semantic vector space locates the similarities, it does not provide a direct means of distinguishing between them. Granted, a great deal of current work is attempting to extract more structure from immense computed vector spaces. But the price to pay is the need for new learning networks, often supervised ones that must be revised and adapted to the linguistic characteristic under study. As a result, the ability of these systems to use computed vectors to account for variations in the meaning of a lexeme in context, has been examined in studies whose results differ significantly in quality [18, 19]. Moreover, to my knowledge, there are no reliable results on opposites, on the possible asymmetry of paradigmatic links as mentioned in the preceding chapter, or on an organization around prototypical semantic meanings. None of these language characteristics have been addressed in current models.

2.2.3 Might Usage and Syntagmatic Agreement Give Exhaustive Access to Content Forms

From the cognitive standpoint, it has been shown that infants make the distinction between living and non-living things, that the processing of these two categories is distributed across different cortical networks, and that the assignment of characteristic features to increasingly specific categories (such as drinking for animals, transporting for vehicles) precede their verbalization[5] [20]. This means that if a concept (for example, *dog*) can be lexicalized and used in various sentences, the link between the lexeme and its content, i.e., its content form, does not result solely from its usage in the language. These experimental findings underline a weak point of approaches in which linguistic usage precedes and determines meaning.

Inversely, it is rather clear that taking into account only paradigmatic relations involving a given lexeme, as proposed in the WordNet project, does not allow one to determine the every-changing set of sentences containing it. Paradigmatic relations alone cannot explain phenomena like meaning creativity in context and more generally of semantic composition.

2.2.4 Lexical Links and Content Form

This chapter was aimed at showing that paradigmatic relations are anchored in content forms and that there are morphisms between the syntactic agreement and the content forms assigned to sentences. I also attempted here to point out the limitations and costliness of approaches founded solely upon these two types of links. For a semantic model and thus a model of content forms, an additional question arises, namely, whether these two types of links could be deemed sufficient even if they were taken as a duality as assumed by structuralists. Over and above any arguments about the necessity of a rigorous definition of a word, or about its substitution by the bundle of utterances containing it, the question that arises concerns whether the language can access its own content and how the content form could make that possible.

Wittgenstein's [21] demonstration using the word *game*, which shows that any attempted definition is partial, also illustrates that a word's content form can never be fully accessed via the language because doing so always requires a supplemental and/or readjustment step. Another illustration: differences or oppositions between the content forms assigned to words are very difficult to state. How, for example, can one verbally characterize the difference between two synonyms such as *origin vs source*, *receive vs get*? Authors of synonym dictionaries [22, 23] have attempted to find an answer to these questions. Although the explanations provided are very useful for speakers, we cannot be certain that they would not be partly invalidated by counterexamples. In the light of the experimental results cited above, it is my conjecture that the content form, that is, the link between an expression form and its content, contains knowledge that is available to our awareness but goes beyond its verbalization. And also that the content form cannot be projected in an exhaustive way onto the axis of language production or of the language's lexical organization. This will bring me to propose a model that relies on both of these types of links, but that constructs the content form from units of another granularity than those of linguistic units. Before that, however, it will be necessary to come back to the questions of grammatical category and function.

---

[5] The children tested in the experiments were under the age of 14 months.

References

1. François-René Chateaubriand. *Mémoires d'outre-tombe*, volume 1. Biblioth que de la Pl iade, 1947
2. Jakobson, R.: Deux aspects du langage et deux types d'aphasie. Essais de linguistique générale **1**, 43–77 (1963)
3. Carota, F., Nili, H., Pulvermüller, F., Kriegeskorte, N.: Distinct Fronto-Temporal Substrates of Distributional and Taxonomic Similarity among Words: Evidence from RSA of BOLD Signals. Neuroimage **224**, 117408 (2021)
4. Louis Hjelmslev. *La catégorie des cas : étude de grammaire générale*, volume 7. Universitets-forlaget, 1935
5. Pustejovsky, J.: The Generative Lexicon. MIT Press, Cambridge, Massachusetts (1998)
6. Antoine Culioli, Catherine Fuchs, and Michel Pêcheux. *Considérations théoriques à propos du traitement formel du langage: tentative d'application au problème des déterminants*. Association Jean-Favard pour le développement de la linguistique quantitative, 1970
7. Lucien Tesnière. 19662. Éléments de syntaxe structurale. *Paris: C. Klincksieck*, 1959
8. Thom, R.: Modèles mathématiques de la morphogénèse. Christian Bourgeois Editeur, Paris (1980)
9. Collins, A.M., Quillian, M.R.: Does Category Size Affect Categorization Time? J. Verbal Learn. Verbal Behav. **8**, 240–248 (1969)
10. Elizabeth K Warrington and Tim Shallice. Category Specific Semantic Impairments. *Brain*, 107(3):829–853, 1984
11. Rips, L.J., Shoben, E.J., Smith, E.E.: Semantic Distance and the Verification of Semantic Relations. J. Verbal Learn. Verbal Behav. **81**(12), 1–20 (1973)
12. Rosch, E., Mervis, C.: Family Resemblances: Studies in the Internal Structure of Categories. Cogn. Psychol. **7**, 573–605 (1975)
13. Aristote. Catégories : De l'interprétation. *Organon*, 1984
14. Fellbaum, C. (ed.): Wordnet. An Electronic Lexical Database. MIT Press, Cambridge, Massachusetts (1998)
15. Kleiber, G.: La sémantique du prototype. PUF, Paris (1990)
16. Thomas K. Landauer, P. Foltz, and D. Laham. Introduction to Latent Semantic Analysis. *Discourse Processes*, 25:259–284, 1998
17. Burgess, C., Lund, K.: Modelling Parsing Constraints with High-Dimensional Context Space. Lang. Cognit. Process. **12**, 177–210 (1997)
18. David Yenicelik, Florian Schmidt, and Yannic Kilcher. How Does BERT Capture Semantics? A Closer Look at Polysemous Words. In *Proceedings of the Third BlackboxNLP Workshop on Analyzing and Interpreting Neural Networks for NLP*, pages 156–162, 2020
19. Aina Garí Soler and Marianna Apidianaki: Let's Play Mono-Poly: BERT Can Reveal Words' Polysemy Level and Partitionability into Senses. Transactions of the Association for Computational Linguistics **9**, 825–844 (2021)
20. Jean M Mandler. On the Foundations of the Semantic System. In *Category Specificity in Brain and Mind*, pages 337–349. Psychology Press, 2005
21. Ludwig Wittgenstein. *Philosophical Investigations*. Blackwell Publishing, 1953
22. Pierre Benjamin Lafaye. *Dictionnaire des synonymes de la langue française: avec une introduction sur la théorie des synonymes*. Hachette, 1858
23. François Guizot. *Nouveau dictionnaire universel des synonymes de la langue française*. Didier, 1861
24. Aristotle. *Metaphysics*. Perseus Digital Library, 1987

Chapter 3
On the Lexical Anchoring of Syntactic Nature and Role

3.1 Structuralism, Syntagmatic Agreement, and Grammar

In his Course in General Linguistics, Saussure presents a critique of traditional grammar[1] as a set of normative—and thus prescriptive—rules, while in a scientific approach it is the facts that must be considered. In his description of the foundations of the structural approach, the distinction between syntagmatic and associative (paradigmatic) relations precedes grammar and provides it with an explanatory framework:"Every syntactic fact belongs to the syntagmatic class" [1]. Accordingly, the categories (noun, adjective, etc.), also called the parts of speech, are derived from a thorough examination of associations (paradigms) and are connected by the feeling of a common value that dictates an identical usage. The syntagmatic relationship, and the word order of the sentence upon which it depends, would be the source of syntax also, and not the other way around, unlike generative grammars, which see syntax as coming first, and being instantiated in the surface structure and ultimately in the lexicon or even totally void of any lexical trace.

The structuralist approach, which starts from the materiality of a language to describe it scientifically by applying contrasting operations along the paradigmatic and syntagmatic axes nevertheless runs up against a methodological flaw, because the expertise needed to carry out these operations depends on assessments made by linguists. Yet examining and evaluating the shared or separate values induced by these relations has never been formalized. What's more, would the examination process be systematic, would it produce the expected more general, more structured, and more abstract view, or would it drown itself in an inconsistent multitude of contrasts?

Based on work in neurophysiology and corpus analysis, I would like to bring up the idea that grammatical notions like part of speech, syntactic role, or even

[1] Saussure defines grammar as including morphology and syntax. I retain this definition here.

Here, I prefer to use the term "syntactic nature" (often called "syntactic category") because it underlines the cognitive anchoring of categories.

S. Ploux, *Semantic Matter and Its States*, Lecture Notes in Morphogenesis,
https://doi.org/10.1007/978-3-032-22809-3_3

sentence organization are not necessarily abstractions emerging from usage, but can find confirmation and content in initially non-linguistic cognitive processes.

3.2 Grammatical Categories

The various categories of lexemes, here the parts of speech, are introduced differently by linguists, depending on whether they situate themselves within a distributional paradigm or a generative paradigm. In the former approach, as mentioned earlier, they represent very broad paradigms derived from the possible substitution of one lexie for another in sentences. In the latter approach, they would be determined by examining the respective roles they play in sentences [2]. However, the notion itself is debated. Some argue that it is difficult to come up with a fixed definition of such categories [3], and that they are more or less marked and distinguished in different languages. Chinese is often cited as a language in which the same word can be a noun or a verb in an undifferentiated way in the same sentence (Maspero [4], p. 36, later taken up by G. Guillaume). Disagreements around these questions accompany the controversy over whether grammatical category is a property of lexemes (*house* is a noun, *to run* is a verb) or, on the contrary, a property that a lexeme takes on in usage. A series of studies rooted in the analysis of the neurological processing of linguistic *stimuli* brings new and extrinsic elements to these debates.

3.2.1 Insights from Neurolinguistics

3.2.1.1 Nouns and Verbs

In a review of the noun *vs* verb dissociation, Vigliocco et al. [5] remind us that certain aphasic patients have more difficulty processing and producing nouns than verbs, while others, greater in number, have the opposite difficulty. It has also been shown that noun processing involves the activation of left temporal lobe areas whereas verb processing is located in areas of the left frontal lobe; for nouns these respective networks are part of larger networks involved in object processing, and for verbs, part of the networks involved in the processing of events or actions. Moreover, verb processing is more time-consuming and requires more costly processing in terms of the areas involved. This greater cost results from the processing load incurred by the actants taking part in the interaction the verb expresses and organizes.

3.2.1.2 Categories and the Lexicon

It should be noted, however, that the vast majority of neurophysiological studies choose stimulus words that maintain congruence between each of these grammatical categories and the prevalent type of associated content: noun-object (e.g., *a ball*), verb-action (e.g., *to run*). To characterize what they believe to pertain specifically to lexical semantics or to grammatical category, and thus to identify the processing specific to the latter, Vigliocco et al. (*ibid*) cited various studies using stimuli that dissociate semantic content and grammatical category. They can be action words, whether verbs or nouns (*to run, a race*) or words deemed to be abstract (*love, to love*). This provides evidence of shared cortical semantic activation irrespective of category. For example—and in line with the studies cited above showing that the neurophysiological processing of a given lexeme involves activations of the cerebral networks specific to its content [6]—there is activation of (frontal) motor areas shared by action words regardless of their grammatical category (*to race vs the race, to dance vs the dance, to jump vs the jump*). Moreover, to dissociate syntactic category and function, the other studies cited present *stimulus* words, either alone or in a sentence context that confers a function upon them. The results obtained seem to show that lexemes are not processed preferentially as nouns or verbs when they are presented alone. Conversely, when in a sentence, they trigger the processing of the grammatical category (or role) they assume therein. In particular, the processing of syntactic integration (and therefore of role) would be carried out for both categories in a common brain area, the left inferior frontal gyrus.

3.2.2 Toward a Reinterpretation of Neurophysiological Results

However, several preconceived ideas found in the neurolinguistic literature raise some questions, namely, the desire to remove any form of semantic content from grammatical categories, and the confusion between syntactic category and syntactic role (which I will return to later). Despite attempts to separate these two notions using words that are not prototypical of a grammatical class but result from the complexification of the language via the application of grammatical operators on contents that are less and less concrete, these authors forget what substantivizing (*to run* → *a run*) or converting a noun to a verb encompasses and involves in the semantics of lexemes. Thus, *a run* is not only the noun format associated with the action of running, but it can also denote its unfolding or its delineation (*the run will begin at 5 o'clock*), and thus a way of transforming a dynamic into an entity and conceiving of it as such.

In spite of this desire to separate semantics and syntax, the results have consistently indicated substantial prototypical anchoring of grammatical categories in content. They are also in line with semantic-content sharing between words, outside of their grammatical category. This point argues in favor of the idea defended here: that semantic matter is malleable in cases of potential changes in the grammatical category of a lexeme in the language, and of the semantic effects produced. Thus, for example, many adjectives and verbs can be substantivized (*red* → *the reds, to play* → *a play*). This phenomenon, called conversion or zero-derivation because the lexeme is unchanged, is known to be massive in English, especially in the case of verb substantivization. It is analogous to derivation via morphological marking of the target category (*judge*, *judging*, *judgment*, etc.). Such conversions or moldings modify the contents in such a way that it becomes consistent with a content form of the grammatical category involved. Substantivizing will tend to reify (*a drink*, *a thing one drinks*; *the reds*, *things that have this feature*), whereas the transformation of a noun into a verb will tend to awaken the inherent dynamics of the thing being designated (*judge* → *to judge*, *monkey* → *to monkey*) by itself or by those it targets (*to bottle* evokes the action of putting something in a bottle). Lastly, note that while the assignment of a grammatical category to a content form is variable and changeable, it is nevertheless grounded in acquaintances with a prototypical category of the content form that determine the limits of the possible changes. Thus, in French, nouns of place that express things that are delineated and spatial (*village*, *cité*, etc.) can hardly be considered verbal in nature. Indeed, these terms are more often circumstants than actants of a process because they cannot readily be a source of interactions *per se*. Sentences in which they act as the subject (*the village burns*, *the village rebels*) contain either an agency-mediated attribution (see below) or a metonymic personification (here, *the village* refers to *the villagers*). And when they are object complements (*discover a village*, *point out a village*), the process that governs them is not specific to them.

3.3 Syntactic Roles

A syntactic role refers to a function in a sentence. For different authors like Tesnière [7], the verb's role is to organize and distribute other roles such as subject and complement.[2] Grammatical category does not determine the role. Accordingly, a noun can be a subject or a complement (*the cat catches the mouse*, *the mouse runs away from the cat*), a verb (in the sense of a grammatical category) can play the role of an actant (e.g., in French "*manger fait grandir*", literally "*to eat makes to grow*", which means "*eating makes one grow*"). Although many neurophysiological studies have found evidence of the participation of the cerebral areas usually involved in semantic composition (left anterior temporal lobe) or as recalled above, in the integration of a sentence's constituents during its processing (inferior frontal gyrus), some data has illustrated that *verbs*, *subjects*, *objects*, and *adjectives* play specific roles.

[2] This statement is in line with Chomsky's proposal [18] that the verb controls the assignment of its arguments' theta roles.

3.3.1 The Verb as the Central Element of Sentence Organization

The syntactic role of the verb—as described by Tesnière who makes it the highest element in the hierarchy that governs its arguments, one of which is the subject—is consistent with the nature of its content, which corresponds to the interaction of (or processing by) actants. In line with this proposal, Ploux and al. [8] conducted a study based on a multivariate analysis of EEG signals (obtained via electroencephalography recordings during the presentation of simple sentences (*the child eats an apple vs the apple the child eats*[3]). This study showed that verb processing takes precedence over that of nouns (in the order of the eigenvector value), and likewise for roles such as subject and complement. These results suggest that in simple sentences containing only a subject, a verb, and a complement, the verb is the primary locus of the syntactic and then semantic processing of the sentence. The finding that the verb plays a predominant role in the sentence is in line with the previously mentioned idea that its more costly processing reflects the integration function it performs.

3.3.2 Subject and Object Roles

Systematic studies of text corpora have revealed that the subjects of sentences are overwhelmingly animate entities (people, animals, etc.) whereas direct-object complements are conversely and also predominantly inanimate entities (for a study on Dutch, see for example [9]). In addition, a study [10] conducted on a wide range of languages showed that for many languages, inanimate entities cannot be the subject of a transitive sentence. Another phenomenon is that the first-person personal pronoun ("I" in English) refers to a person except in cases of fictitious attribution.

Furthermore, according to a longitudinal study of cognitive development in childhood, carried out by Piaget [11], the child, at the onset, would be a subject only, the sole subject and the center of the surrounding environment, which he/she would take in as a whole via the activation of sensorimotor schemas. Then, as the child experiments with the world, it materializes into permanent objects; this account of development makes the subject the primary source of interaction, while the object, seen here as the target of that interaction, becomes a corollary. Next, the child becomes capable of assigning the attributes of the subject—the source of interaction and dynamics—to a target object. Conversely, along with all other constructed subjects—the child would be able to conceive of her/himself as an entity with spatial and functional closure, namely, as an object. In a way here, the cognitive bases of the syntactic roles of subject, verb (interaction), and target object, which will come later in child development, precede the construction of a nature common to the target objet and the subject, just like that of a noun, either the subject (the source of interaction)

[3] In the experiment in French: *l'enfant mange une pomme* vs *la pomme que mange l'enfant*.

or the object (the target of that interaction). What seems to be occurring here is an acqaintanceship, firstly, between the grammatical subject and content forms that are action and agency sources, and secondly, between the object complement and content forms that are extrinsic-agency targets. In a neurophysiological study, Frankland and Greene [12] found evidence of activations of neighboring, but different, brain areas for subject-agent nouns and object complement-patient nouns in sentences, activations that conform to these prototypical associations between grammatical role and actants (for example, *The dog chased the man*). However, to my knowledge, no studies have raised these same questions while separating agency from syntactic role.

Thus, the same acquaintance phenomenon between noun and entity, or verb and action, seems also to occur here between subject and agent, or object and target. But, as mentioned above, the language system is not necessarily bound by prototypical anchoring. We find agency-mediated attributions (examples from French: *la voiture démarre* (the car starts up)) or inversions in passive constructions (*la pomme a été mangée par l'oiseau* (the apple was eaten by the bird)). Such inversions in French can be explained by the conflict between the subject-agent and object-patient acquaintanceships, as well as by theme-rheme polarity, or even the absence of a generic subject marked by the use of a reflexive verb (*ça se voit*, literally, it sees itself) for *on voit ça* (one sees it), or *les fraises se mangent à la dérobée* (literally, strawberries eat themselves on the sly) for *on mange the fraises à la dérobéé* (one eats strawberries on the sly).

3.3.3 The Role of Adjectives

In a series of earlier studies [13] that used semantic vectors computed from co-occurrences in corpora to represent the stimulus words presented to subjects, one study [14] made use of machine learning to decode the brain-activity patterns specific to the processing of each word. The decoding process was then applied to tracking the cerebral activity linked to the semantic and neurophysiological processing of phrases of the adjective-noun type (*tasty tomato*). The results (here, in English where the adjective comes before the noun), showed that adjective-specific brain activity continued during the processing of the noun it qualified. Although this finding seems to indicate concomitant processing of the content forms detected by the vectors of the noun and the adjective modifier, and thus a trace of their respective roles in the construction of meaning, it does not describe the qualifying function of the adjective and its semantic effects on the noun. As suggested by the authors of the study, a comparison with other languages in which the adjective-noun order is reversed, and also with studies contrasting the cerebral processing of other parts of speech, would be useful in gaining a better understanding of the specific role of the adjective, a topic that remains largely unexplored in the literature.

3.4 Syntactic Structure of Sentences

Many studies have pointed out a similarity between the constituent structure of a sentence and the dependency links between those constituents, often presented as a hierarchical structure that unfolds sequentially, as in other types of structures that require coordinated planning like gestures [15], music [16], and prosody [17]. Neuroimaging experiments inquiring into this similarity have provided evidence that the areas involved in sentence-structure processing are shared with those of gestures. In particular, there is substantial documentation of the joint involvement of left Brodmann area 44, an important seat of the mirror neuron system, in the processing of syntax and the organization and planning of gestures—themselves composed of motor subunits—for carrying out tasks like grasping an object. The nature of the sharing is under debate. Are the same mechanisms at work in these cognitive functions or is each function derived from a specific amodal capability to process hierarchical structures that a temporal realization lines up seqentially? Whatever the case may be, and contrary to the idea of the autonomy of syntax, these studies reveal its cognitive anchoring in mechanisms that are not specifically linguistic.

3.5 Revisiting the Predominance of Lexical Relations Over Grammar Among Structuralists

From all of this data, we can draw out the view that grammar does not derive solely from the application, *a posteriori*, of paradigmatic and syntagmatic relations. Indeed, far from resembling abstractions or generalizations, the cognitive correlations identified point instead to anchoring in so-called low-level functions (ability to delineate and discriminate entities or actions, to structure and coordinate constituents, etc.). This leads to the idea that grammatical categories and functions, like the signs of structural linguistics, are composite elements that link up regularities inscribed in expression forms (such as morphological markers) to regularities of content. In a way, this point of view systematizes, for grammar, the expression-form *vs* content-form duality initially elaborated for the sign. Drawing on these comparisons between neurolinguistic findings, linguistic facts, data, and theories, along with the interpretation of their coherence, I will attempt in what follows to compare and contrast the main models in computational semantics, both current and past.

References

1. de Saussure, F.: Course in General Linguistics. Philosophical Library, New York (1959)
2. Emonds, J.: Parts of speech in generative grammar. Linguist. Anal. **7**(1–2), 3–42 (1987)

3. Auroux, S.: La grammaire générale et les fondements philosophiques des classements de mots. Langages **92**, 79–91 (1988)
4. Maspero, H.: La langue chinoise. Conférences de l'Institut de Linguistique de l'Universit de Paris, année **1933**, 33–70 (1934)
5. Vigliocco, G., Vinson, D.P., Druks, J., Barber, H., Cappa, S.F.: Nouns and verbs in the brain: a review of behavioural, electrophysiological, neuropsychological and imaging studies. Neurosci. & Biobehav. Rev. **35**(3), 407–426 (2011)
6. Hauk, O., Johnsrude, I., Pulvermüller, F.: Somatotopic representation of action words in human motor and premotor cortex. Neuron **39**(41), 301–307 (2004)
7. Tesnière, L.: 19662. Éléments de syntaxe structurale. C. Klincksieck, Paris (1959)
8. Ploux, S., Déprez, V.: Organization and independence or interdependence? Study of the neurophysiological dynamics of syntactic and semantic processing (2018). arXiv:1804.05686
9. Bouma, G.J.: Starting a sentence in Dutch: a corpus study of subject-and object-fronting. Ph.D. thesis (2008)
10. Fauconnier, S.: Differential agent marking and animacy. Lingua **121**(3), 533–547 (2011)
11. Piaget, J., Chomsky, N., Piatelli-Palmarini, M.: Théories du langage, théories de l'apprentissageâŁ¯: le débat entre Jean Piaget et Noam Chomsky. Éditions du seuil, Paris (1979)
12. Frankland, S.M., Greene, J.D.: An architecture for encoding sentence meaning in left mid-superior temporal cortex. Proc. Natl. Acad. Sci. **112**(37), 11732–11737 (2015)
13. Mitchell, T., Shinkareva, S., Carlson, A., Chang, K.-M., Malave, V., Mason, R., Just, M.: Predicting human brain activity associated with the meanings of nouns. Science **320**, 1191–1195 (2008)
14. Fyshe, A., Sudre, G., Wehbe, L., Rafidi, N., Mitchell, T.M.: The lexical semantics of adjective–noun phrases in the human brain. Hum. Brain Mapp. **40**(15), 4457–4469 (2019)
15. Kemmerer, D.: Revisiting the relation between syntax, action, and left BA44. Front. Hum, Neurosci (2022)
16. Patel, A.D.: Language, music, syntax and the brain. Nat. Neurosci. **6**(7), 674–681 (2003)
17. Heard, M., Lee, Y.S.: Shared neural resources of rhythm and syntax: an ALE meta–analysis. Neuropsychol. **137**, 107284 (2020)
18. Chomsky, N.: Lectures on Government and Binding: The Pisa Lectures, no. 9. Walter de Gruyter (1993)

Part II
Language Unit *vs* Model Unit

"Language can only be explained by language" (Hjelmslev 1935, p. 135)

Chapter 4
Models and Domains: Effects and Fit

This chapter deals with the fit between a model and the domain being modeled. I will attempt to show that any attempt to model not only requires applying an adjustment process to the domain, but can lead to formal research. I will also introduce the idea that the level of the objects and phenomena in the domain is not necessarily the best level of granularity for the model's entities. This second point will be developed in greater detail in the section on polysemy.

4.1 Mathematical Physics, a Widespread Example

While the terms "to model" and "modeling" are recent, having appeared in the late 20th century,[1] the search for a mathematics of the sciences is much older. In investigations of this type, the mathematizing of mechanics and later that of the physical sciences generally served as a reference. The abstraction process and the advancements thus obtained were the fruit of a lengthy and thorough process of the adaptation of mathematics to physical phenomena, which involved both the creation of formal notions and the redefinition of the object of study. This is what Alexandre Koyré [1] showed in his analysis of the birth of modern mechanics, which overthrew the view of a closed hierarchical world composed of natural objects, in favor of a view of the world as uniform and infinite.

For Aristotle, the understanding of physical phenomena rests on a description of things, their properties, and their own qualities. And in this view, movements are properties of each and every thing. Movements include not only translational movements (bodily locomotion), but also internal movements (like growth). "There

[1] The scientific meaning of theses terms in English dates back to the 20th century (OED).

S. Ploux, *Semantic Matter and Its States*, Lecture Notes in Morphogenesis,
https://doi.org/10.1007/978-3-032-22809-3_4

is no such thing as movement outside of things [...] There are as many kinds of movement and change as there are kinds of beings [...] Development and reduction are movements of beings that are developing, or on the contrary, being reduced; for here, there is no common expression for these two ideas; generation and destruction are movements of beings that are generated and are being destroyed, just as translation is the movement of beings that are transferred." ([2], Book III; our translation).

The development of modern mechanics separated these natural entities and built a space whose entities are not physical objects as much as they are abstractions, such as a point, speed and acceleration, which are described as the first and second derivatives of the series of points travelled. Newton [3] stated: "To the extent that it appertains to the mathematical sciences, it is reasonable that I largely abstract it from physical considerations." The movements of bodies, i.e., their initial positions and their trajectories in this new paradigm are geometric objects. So this amounts to a narrowing of the notion de motion, which, as such, is applied to objects that lack an essence and properties of their own and are subjected to forces or fields, like gravity, that are not individualized. This abstraction and change in level and scale (we are going from observable, natural objects to infinitely small entities) was accompanied by the appearance of mathematical notions such as a uniform, infinitesimal space, and of a new formal framework, namely, differential and integral calculus. The accordance between the mathematical and physical sciences, both reductionist and productive, enabled the computation of the equilibrium and movements of bodies subjected to specific initial conditions, thus providing answers to questions that could not be answered in investigations based on descriptions and classifications of phenomena.

The example of modern mechanics (simply skimmed over here) gave rise to the idea that, if they are properly adapted, mathematical models can provide theories with a basis for synthesizing, analyzing, representing, and in some cases simulating and computing entities, facts or processes. Moreover, since the latter half of the nineteenth century, approaches to formalizing were taken up in an attempt to obtain similar effects in the human sciences. Models were proposed in anthropology and in the field of cognition. Boole defined an algebra for the laws of thought [4]; Levi-Strauss and Weil relied on the abstract notion of Klein group (in the algebraic sense of the term) in order to account for kinship systems in anthropology [5]. But it is more recently, in linguistics, that the most extensive attempt to formalize took place. Here, I propose a brief analysis of the formal choices made with respect to the two criteria stressed here: delineation of the domain under study and the fact that domain units and model units become one and the same thing.

4.2 What Model for What Linguistic Theory?

4.2.1 Discrete and Continuous as Catalyzers of Linguistic Controversies

At the roots of the debate. Endeavors to formalize have mainly looked at existing paradigms and how they could be taken into account in linguistic studies and traditions. In this context, this discrete vs continuous duality was the principal formal factor that fed the various debates and controversies. Remember that the dissent was rooted in an earlier one that opposed Russell's logistic approach, wherein pure mathematics was taken to be logic, and a view of mathematics as a field in its own right (not purely logical) defended by geometricians. Accordingly, Poincaré proposed the idea that sensible experience and intuition give rise to mathematical discoveries, a position explained in terms of his definition of continuity in mathematics as being drawn from sensible experience [6]. The tension between an often discrete, purely logical mathematics and dynamic systems, was long-lasting. We find it even in Thom's work.

Linguistic Arguments. The idea expressed here by Russell, that logic is to philosophy what mathematics is to physics, went beyond philosophy. Similar ideas were already present in Boolean algebra and in Frege's [7] first-order predicate logic applied to the meaning of utterances. The formalization of logic is in line with the ideas of language autonomy, symbolic and discrete representation, and the notion of generativity. In contrast, the continuous paradigm provides support not only to opponents of the modularity of language and its components (syntactic, semantic, etc.) but also to those who defend the intrinsic variability of the meaning of lexical units and the dynamic construction of meaning in context. Related to controversies are various questions about the delineation of the domain of linguistics. Where should language be situated among the set of cognitive abilities? Is language essentially syntax as proposed by Chomsky, for whom the relation between sound and meaning is determined by syntactic structure? Is semantics cognitive or logical? Are semantics and syntax separable or inherent in each other? As for lexical semantics, is it conceptual and derived from a logical system of necessary and sufficient conditions or is it anchored in usage and drawn from the nature of notions like family resemblance? This last question was addressed in the preceding chapter from the angle of paradigmatic *vs* syntagmatic relations.

4.2.2 Logic, Language Autonomy, and the Generative Theory

Impact of the development of modern logic and its appeal for studying language. Chomsky strove to construct a scientific and formal approach to linguistics and linguistic competence in order to go beyond an "entomological" type of approach to language events, and to lay the groundwork for a universal explanatory framework.

Thus, the approach of logicians was adopted to define an axiomatic of grammar based on the chosen language's entities (sentence (S), noun phrase (NP), verb phrase (VP), noun (N), verb (V), determiner (D), etc.) and its lexicon, and of rewrite and transformation rules. Each rule is represented either as a derivation (for example: $S \rightarrow NP, VP; NP \rightarrow D, N$) or as a tree structure showing the steps in which the rules were applied and the hierarchical prevalence links between the constituents. The purpose is to be able to generate all well-formed sentences—and only those sentences—using an approach analogous to logical inference.

In the same formal trend, traditional so-called Aristotelian psycholinguistics, followed by lexical ontologies, attempted to define a concept by a list of necessary and sufficient properties, as recalled above. Thus, a concept would be a logical form and an *item*'s membership in the corresponding category would be a formal demonstration derived from applying the definition and the inherited properties via syllogisms (for example: All birds have feathers. A sparrow is a bird. Therefore, a sparrow has feathers).

In some ways, the logical approach to linguistics resembles Russell's logical approach to philosophy or mathematics. The idea is to use the formalism of logic to ground the autonomous nature that the linguistic theory claims to have. Here, I will address the influence of this paradigm in terms of the notions of generation, autonomy, and questions regarding its physiological embodiment.

Control or generation? Based on the idea that sensitive experience and intuition are necessarily involved, Poincaré [6, 8] argued that formal logic is not the sole bearer of the mathematical results and knowledge. Logic would make it possible to control reasoning and thus, to avoid erroneous thinking. Opposed to this viewpoint, and in accordance with the logistic paradigm, Chomsky's notion of generation wherein formal rules are applied, is a production system. This raises two interrelated questions. Firstly, while axiomatic systems are indeed conveyed by a language that is formal, granted, but that rests on the ability to express reasoning in words, do such systems make language into a self-generated system? Secondly, is the vocabulary level of an axiomatic system the most suitable one for formalizing language? The latter question will be addressed in the next chapter.

Formal self-sufficiency and autonomy of language. The formal logical framework in question here involves a closed conception of language. The set of all well-formed sentences, although potentially infinite, is determined from a predefined set of vocabulary words and the rules of an axiomatic system. All the necessary things are present. The object of study—language—is itself delineated by this generative capacity, which informs or makes the connection between two distant levels of the syntactic core, phonetics and semantics. Language becomes a closed system of rules, which in this formal framework, defines its nature. Chomsky pushes this formal autonomy to the point of making it an embodied language organ [9].

Autonomy hypothesis and innateness. Going from abstract autonomy to biological modularity requires empirical validation, given that Chomsky defends language modularity via its innate character that is analogous to that of all other organs (eyes, heart, etc.) and whose development is genetically inscribed. However, it remains to be explained how the links between the genome and the formation of the language module and its structure are determined. This would require grounding the generative rule system elsewhere than in syntactic structures. And that would open up the model to units of another level, thereby partially invalidating the autonomy and modularity hypotheses. Besides, these links between the genome and linguistic structures and functions have yet to be fully traced and explicitly stated. Current results have pointed out correlations between certain genes and the brain locations involved in human language functions [10], without, however, describing the set of processes underlying the construction of the linkage between genome and the language faculty.

4.2.3 The Continuous Paradigm

The advocates of the continuous paradigm present a series of linguistic arguments against logical, discrete models. First, the extralinguistic parameters upon which language is based, such as perception, movement, pragmatic context variations and so on, are continuous in nature, a nature that is thought to have linguistic repercussions. Second, the polysemy and intrinsic variability of word meaning are better represented in a continuous space. Third, difficulties encountered in accounting for the findings of psycholinguistic experiments on categorization (and more particularly on prototypes) in a purely logical framework have made it necessary to rely on a paradigm that involves either weighting (which uses values taken from real numbers as in fuzzy logic) or gradients in a continuous space. Fourth, continuity would allow a large set of data to be synthesized. This last argument is methodological and does not presuppose that a continuum is used to adapt to the intrinsic nature of linguistic units and processes. While the oppositions to discrete modeling are multiple, these objections, as a whole, nevertheless shift the focus of language theories from regularities or deep structures of expression forms, to those of content forms or even more directly of content (Wittgenstein's "what I am talking about", see footnote[2]).

Listed here are some standard examples of continuous formal frameworks: vectorial, dynamic, and geometric. This classification is artificial from a mathematic standpoint, but I use it here because it corresponds to the different types of research dealing with linguistic modeling. Also, each one has been brought to bear, depending on which one of the above arguments is being defended. Note that fuzzy logic—which was seen as a solution for reducing the rigidity of discrete logics—keeps the theoretical hypotheses intact since is consists of integrating real coefficients as weights for making inferences or applying derivations. For this reason, it will not be discussed here.

[2] When I give the description: 'The ground was quite covered with plants' do you want to say that I don't know what I am talking about until I can give a definition of a plant?" [21].

4.2.4 Vector Space, Similarity, and Semantics

The notions of family resemblance, gradient, and similarity previously introduced as alternatives to purely logical and definitional approaches in lexical semantics and psycholinguistics of prototypes are echoed in the notion of neighborhood, itself subtended by that of distance in a vector space. As mentioned, the coordinate-based uniform space was proposed in the seventeenth century by Descartes and Fermat. It was reworked and enriched over the subsequent centuries through the integration of algebraic structures (among others) like that of field [11]. Accordingly, once a vector is assigned to each linguistic unit considered, the algebraic properties of fields allow one to represent the semantic proximity (or vector distance) between two of these units, their semantic composition within a phrase or sentence through their linear combination [12], or their difference taken as a semantic feature that separates them. Mikolov [13], for example, showed that differences between the computed vectors assigned to pairs like *king-queen* or *man-woman* are relatively constant. This has been interpreted to indicate that it is possible to compute a vector characterizing gender.

Another founding argument set forth in the vectorial paradigm is that meaning cannot be described solely by a definition and is drawn from usage. In this view, vectors representing meaning are computed by scanning text corpora (usage) and are based on a count of co-occurrences within potentially sliding windows (*n* words), sentences, paragraphs, or documents.

4.2.4.1 Usage *vs* Syntax

The methods used to compute vectors assigned to language units via corpus scanning have led to great advances in automatic processing. Partly to support the results obtained and partly by theoretical ambition, some have argued that sentence structure is neither necessary nor primary. Language would therefore be nothing more than learned statistical regularities. Granted, certain ways of building vectors take the order of linguistic units in the chosen window of occurrences into account: this is true of the system *Hyperspace Analog to Language System* (HAL, [14]) and of certain recurrent neural networks. The sequence is a minimal form of structure. It remains, nonetheless, that apart from taking into account a sequential order, syntactic structures and functions are omitted from the computation. Studies that do not ignore syntax seek to find traces of it *a posteriori* in vector configurations, as recalled above.

This is a reversal of the generative approach, which postulates the existence of a syntax module that is deep, innate, and takes precedence over lexical semantics, thought to be contingency-based. Following Chomsky, the argument of *stimulus* paucity in childhood has often been called upon by generativists as arguing in favor of the innateness and primacy of syntax rather than usage and learning theories as a language acquisition framework. However, this argument has been partially invalidated by studies showing that children are subjected to a much larger flow of language *input* than the generativists thought. The fact remains that the quantity of input needed to feed current neural networks and give them generative capabilities (such as question-answer systems, translation, summarizing etc.) is nonetheless massively larger than that which any young child might hear. From this angle, neither usage-based approaches supported by vectors, nor generativist models supported by logic, have closed the debate on the deep nature of language.

4.2.5 Dynamic Systems, Verbal Interaction, and Meaning-Construction Processes

In the face of these last two highly prevalent trends, each grounded in its own manner on language regularities, the theory of enunciative operations [15] stands out from others by the fact that its objective is to study language activity without masking the phenomena of variability.

4.2.5.1 Accounting for Variability

Culioli defends the idea that linguistic phenomena form dynamic systems that are regular, but involve a margin of variation. They are both stable and malleable. For a given linguistic unit, this means revealing a schematic form that describes its invariant part and the variable part that governs, for example, the unit's polysemy and the conditions that affect its deformability. These conditions are based on a wide variety of factors: the co-text in which the unit is employed, the units the co-text contains, their syntactico-semantic properties, properties, which cannot be classified as either syntactic or semantic, and so on. The formal notions brought to bear to characterize the schematic forms are borrowed from mathematical topology.

A few studies have promoted the use of topological notions in an attempt to formalize the dynamic construction of a lexical unit's meaning as a function of its schematic form and of variations in its co-text of use, and thus to formulate a linguistic theory in the field of dynamic systems. However, this endeavor could not be generalized to all language units, since a given verb or adverb studied alone requires manually determining a schematic form along with its invariant and deformation parameters [16]. This would explain why the project was not extended beyond the examples studied.

4.2.5.2 Dynamic Systems and Roles in Verbal Interaction

The most highly developed way of relying on dynamic systems is not derived directly from the theory of enunciative operations but from Thom's proposal to model phenomena around the notion of universal unfolding and theorems related to it. Universal unfolding involves normalizing and drawing up an inventory of all forms resulting from a perturbation of singular points. The perturbation requires modifying the values of some of the control parameters. The number of parameters that must be modified depends on the "depth" of the singular point and the dimension of the spaces at play. From the ability of the theory to find and describe the possible outcomes of a mathematical function around a singular point, Thom devised a set of models to account for morphogenesis phenomena, principally in the biological sciences but also in the human sciences. In this framework, verbs are associated with a dynamic that is organized around a singular point and represents an interaction between protagonists (subject and complements) (see the example of the French verb *couper* (to cut), Fig. 2.2). This model sheds light on the difference between the nature of the verb and the actants in the sentence. The verb is a time-based dynamic of interaction; the actants are represented as balls in a multidimensional space (like bodies in physics) and can be likened to points in a space.

However, from a linguistic point of view, several objections can be raised. The first is that unlike the theory of enunciative operations, this model is more a model of content than of the content form. The French verb *couper* has many senses, and the link between its expression form and all of the contents attached to it goes beyond and encompasses interactions other than those represented par the proposed schema. For example, in the expression *couper du vin avec de l'eau* (literally, "cut wine with water"), the wine is not separated but diluted. The second objection follows from the fact that since nouns, the protagonists of the interaction, are represented by points, i.e., atoms without parts, the model cannot adapt to the complexity of their content form, whereas in the language, the meaning of the verb (here *couper*), will depend on the meanings of the subject and the objects. This is the case, for example, in *couper la parole* (*interrupt someone's speech*), which marks an interruption where the meaning of "speech" is associated with a flow or rate (produced by a person) and not with any other sense of *parole* (speech) (the shared ability involved in the expression *l'apprentissage de la parole* (learning to speak)). We find the same co-determination phenomenon for the verb *couper* and the flow or rate of its complement, in the expression *couper les vivres* (cut off the supplies).

Yet another objection is that the number of verbs in a given language (several thousands in French) far exceeds the size of the typology of interactions given by this mathematical framework (i.e., a few dozen). It seems, then, that in its current state, this typology has a level of granularity that is too coarse for the semantic modeling of verbs. Finally, Thom argues that subjects and objects are trivialized in the notion of protagonist, and, as we have just seen, are identically represented as points. This position conflicts with Thom's own idea that the three layers of embryogenesis—ectoderm, mesoderm, and endoderm—are destined to become the epidermis and the nervous system (ectoderm), the muscles, skeleton, and blood vessels (mesoderm), and the digestive system (endoderm). These layers would correspond to the respective biological anchorings of the syntactic functions "subject", "verb", and "object" [17]. As a result, each layer (syntactic function) should acquire a differentiated nature in this analogy. This runs counter to the amalgamation of subjects and objects in the verb morphologies typically proposed.

I might also mention Gärdenfors's [18] later work proposed in opposition to symbolic paradigms. It seeks to link words to their conceptual content rooted in our sensible experiences. In this framework, content is modeled according to the type of sensory input (color, shape, texture, taste, etc.) by way of geometric spaces representing them, e.g., for color, it is the color wheel. Concepts are represented by sets of convex regions included in the geometric spaces associated with each type of sensory input. The syntactic categories are projected differently onto these geometric spaces. Adjectives are projected onto a simple convex geometric region. For example, *hot* refers to a temperature range that varies as a function of the word being qualified. In "a hot day" the range associated to *hot* would be between 25 and 35 °C; in "a hot tea", it would be between 40 and 60° (these figures are given here for illustrative purposes only). Nouns, like concepts, are clusters of convex regions. The word *apple* would be represented by the following cluster: the subregion of the color wheel corresponding to the possible color of an apple, a specific range of acidity for taste, and so on. Verbs are characterized by a dynamic. Gärdenfors's proposal is to represent verbs by patterns of forces (rather than by geometric patterns) and by differential equations that link the different actants.

4.2.5.3 Dynamic Systems: Model or Metaphor?

Dynamic systems provided an answer not only to the desire to get closer to linguistic variability phenomena like polysemy, but also to ground sentence structure in major biological functions, thereby underlining its semantic facet. They make it possible to break down the boundaries between syntax and semantics by having syntactic form be the counterpart of actancial interactions, and more specifically grammatical-case-based ones [19]. However, in this explanatory approach, each lexeme, verbal schema, preposition, etc. requires a targeted step performed by an expert, an approach that has not yet been materialized in a large scale computer implementation. One aspect that could limit the systemic scope of this paradigm is that computers are binary machines and only approach the continuous by numerical approximation. They are not machines for processing singularities.

4.2.6 Geometry and Content Forms

A final approach consists of no longer using vectors to represent linguistic units themselves, but raher their various senses, and then computing the semantic organization of lexical units from the variety (or region) that encompasses these sense vectors. This is the case of the Semantic Atlas model and the ELMO and BERT systems, which compute context vectors. In the next chapter, I will describe the characteristics of the above models, not in terms of the delineation of the linguistic domain, which remains close to the domains of other continuous approaches, but in terms of the dissociation between the model's units and the domain's units, a strong characteristic of mathematical physics as mentioned at the beginning of this chapter.

In sum, as in the case of the emergence of classical mechanics, the object of study often gets redefined. In discrete approaches, it is the acquaintanceship between logic and the syntactic component of language that prevails. This acquaintanceship is not specific to this particular domain, insofar as logic is itself likened to a form of mathematical syntax applicable to science in general [20]. Limiting linguistics strictly to the domain constrained by this acquaintanceship results in marginalization of semantics, or at best an attempt to reduce it to a logical form and a derivation or inference system [21]. On the other hand, what takes precedence is the acquaintanceship between the continuous and everything which, in the link between expression form and content form, is part of a process, whether a process of genesis, or of meaning construction and variation. For certain authors, syntax is a set of abstract schematic forms derived from a substratum of continuous regularities; for others it cannot be separated from semantics.

References

1. Koyré, A.: Du monde clos à l'univers infini. vol. 301. Presses universitaires de France (1962)
2. Saint-Hilaire, B.J., et al.: Physique d'Aristote ou Leçons sur les principes généreaux de la nature. vol. 2. A. Durand (1862)
3. Newton, I.: De la gravitation, suivi de Du mouvement des corps. Gallimard (1995)
4. Boole, G.: An Investigation of the Laws of Thought: on which are Founded the Mathematical Theories of Logic and Probabilities. vol. 2. Walton and Maberly (1854)
5. Lévi-Strauss, C.: Les structures élémentaires de la parenté. vol. 2. Walter de Gruyter GmbH & Co KG (2020)
6. Poincaré, H.: Le continu mathématique. Revue de Métaphysique et de Morale **1**(1), 26–34 (1893)
7. Frege, G.: Sens et dénotation. Écrits logiques et philosophiques, pp. 102–126 (1971)
8. Poincaré, H.: La Science et l'hypothèse. Flammarion (1968)
9. Piaget, J., Chomsky, N., Piatelli-Palmarini, M.: Théories du langage, théories de l'apprentissage: le débat entre Jean Piaget et Noam Chomsky. Éditions du seuil, Paris (1979)
10. Mekki, Y., Guillemot, V., Lemaître, H., Carrión-Castillo, A., Forkel, S., Frouin, V., Philippe, C.: The genetic architecture of language functional connectivity. Neuroimage **249**, 118795 (2022)
11. Peano, G.: Calcolo geometrico secondo l'Ausdehnungslehre di H. Grassmann: preceduto dalla operazioni della logica deduttiva. vol. 3. Fratelli Bocca (1888)
12. Kintsch, W.: Predication. Cogn. Sci. **25**(2), 173–202 (2001)
13. Mikolov, T., Sutskever, I., Chen, K., Corrado, G., Dean, J.: Distributed representations of words and phrases and their compositionality. Adv. Neural Inf. Process. Syst. 3111–3119 (2013)
14. Lund, K., Burgess, C.: Producing high-dimensional semantic spaces from lexical co-occurrence. Behav. Res. Methods Instrum. Comput. **28**, 203–208 (1996)
15. Culioli, A.: Pour une linguistique de l'énonciation: opérations et représentations. Editions Ophrys (1990)
16. Victorri, B., Fuchs, C.: Polysémie et construction dynamique du sens. Hermès, Paris (1996)
17. Thom, R.: Modèles mathématiques de la morphogénèse. Christian Bourgeois Editeur, Paris (1980)
18. Gärdenfors, P.: Conceptual Spaces, the Geometry of Thought. MIT Press, Cambridge, Massachusetts (2000)

19. Petitot-Cocorda, J.: Morphogenèse du sens. Presses Universitaires de France, Paris (1985)
20. Cavaillès, J.: Sur la logique et la théorie de la science. Vrin (1997)
21. Pustejovsky, J.: The Generative Lexicon. MIT Press, Cambridge, Massachusetts (1998)

Chapter 5
Units of Meaning, Model Units, and Polysemy

5.1 Distance Between Units of Meaning and Model Units, or Lack Thereof

In the preceding chapter, I established a disjunction between domain units and model units during the construction of mathematical mechanics. A physical body is located in a uniform and infinite space, itself composed of infinitesimal points without content. The physical body is not defined by these infinitesimal units. Yet in the search for a mathematics suited to linguistics, the various approaches mentioned above were unable or only partly able to separate them.

In the logic-based paradigm, for example, language entities are represented by symbols, i.e., the units of the formal model proposed to describe them. To the units "sentence", "verb phrase", "noun phrase", "verb", "noun", and "determiner", Chomsky's initial systems of rewriting rules or derivation rules assigned variable names that differed simply by convention (S, P, NP, V, N, Det, and so on) and by the rules in which they played a role. In the semantic domain, first-order logic (inherited from Frege) separated predicates and the entities of the interpretation domain (objects in the world, etc.). This separation was reused in other semantic models such as Montague's, and in the so-called dynamic semantic theories such as discourse representation theory (DRT) Kamp [1], and segmented discourse representation theory (SDRT) [2], whose field of study extends from clauses to discourse. The predicate, itself a symbol, is a model unit directly representing a verb or a property. For instance, the clause $friend(x, y)$ would be a reformulation of the sentence *y is a friend of x*. Thus, while both of these types of elements are distinguished in terms of the way they are rewritten in the model, each one receives a symbol that represents it and is definitively assigned to it.

Furthermore, in tree-structure representations, which often accompany syntactic-derivation rules, the resemblance to formal units is even more obvious. Each entity becomes a node in the tree and the different types of linguistic links between nodes are reduced to edges. This causes theories like government and binding to be exposed

S. Ploux, *Semantic Matter and Its States*, Lecture Notes in Morphogenesis,
https://doi.org/10.1007/978-3-032-22809-3_5

to, and accompanied by, a tree structure, even though the scientific content is not fully inscribed therein. For example, although the governing elements like the verb are indeed located higher up in the syntactic tree, to indicate their predominance, the hypothesis that the verb selects the subject of the sentence is not marked in this formal representation composed solely of nodes and directional branches. And in general, the differences in the nature and function of the parts of speech cannot be derived from this homogeneous and totally static representation itself. Likewise, transformations such as the possibility of going from an active form to a passive form are reduced to modifications within the syntactic tree, achieved by moving the nodes and sub-trees within the tree. In the literature, these tree modifications are described by relying on linguistic arguments and are done *by hand*. Thus, this way of symbolizing domain entities, which equates an entity in the domain with an entity in the model without differentiating them, amalgamates the domain units and deprives them of their own characteristics, to the benefit of a uniform model.

Vector models and logical models share this non-differentiation between the domain level and the model level: each entity in the language is associated with a fixed vector. As in the symbolic approaches, this representational homogeneity (everything is a vector) conflicts with what linguistics and neurolinguistics have revealed. For, as recalled above, if a verb is not a noun and a word is not a sentence, assigning them formal units of the same nature obviously deprives the model of the possibility of accounting for their intrinsic differences. And the possibility of constructing classifiers capable, *a posteriori*, of separating the different parts of speech on the basis of co-occurrence profiles in sentences does not imply that the model possesses a distinctive representation of their nature and, all the less so, of their own specific function.

Lastly, in Thom's dynamic paradigm, a verb is an actantial schema when—in the same manner as in physics—its arguments are bodies likened to points moving in an abstract space governed by the interaction dynamics. Here then, the verb is represented by a dynamic depicting the process that unfolds in a space whose units are incommensurate with it. Recall, however, that this type of modeling, which involves differentiation of nature and function between the process and the actants, is more like a model of content than of the content form because the polysemy of the process and the actants, and its effect on the interaction schema, are not taken into account.

In sum, when it comes to logic-based and vector-based approaches, the fact of associating a model unit to a language unit, and the fact that there are no other types of model units than those attributable to a language unit, brings us back to the aforementioned issue of self-generation and modularity. Indeed, in these paradigms, there is no extrinsic link between the language and any other level of representation. Well-formed sentences in a logic-based model are the result of rules applied to linguistic elements such as the lexicon and syntactic variables. In the vector-based framework, semantic-composition operations can be reduced to combinations of linguistic vector entities, which also places this approach in a self-generation framework. I have shown how such intra-linguistic generation conflicts with innateness hypotheses and neuroimaging data. I propose here to point out the limitations of this framework when it comes to addressing a specifically linguistic phenomenon, namely, polysemy. Studying polysemy requires defining units of meaning that pose two major problems. Firstly, their perimeter, as we shall see, is a source of debate, and secondly, the linguistic units upon which they are based are not constructed on the sole basis of expression forms (words, parts of speech, sentences, etc.).

5.2 The Choice and Definition of Semantic Units in the Light of the Phenomenon of Polysemy

In Chap. 1, I attempted to show that the context-based variations in meaning of a lexie and the extent of its content form, are mutually inherent. I will begin here by recalling that the very existence of the phenomenon of polysemy remains contested. While referring to the various controversies, I will introduce the choices made to represent semantic units in the different formal proposals.

5.2.1 Polysemy: For or Against

Dictionaries and Drawing Up an Inventory of Senses

It is most likely in single-language dictionaries—the first of which described European languages dating back to the seventeenth century [3]—and thanks to the expertise of the lexicographers who designed and produced them, that the phenomenon of polysemy is best illustrated. Indeed, in this type of dictionary, each entry is organized according to the parts of speech involved and its different senses. Each subdivision may be accompanied by a definition or gloss, examples of characteristic uses, or sometimes an etymological explanation of the meaning, and can be presented as a list or as a hierarchical classification. For the French word *défendre*, for example, the various dictionaries propose different structures. Le Robert separates the word into two main senses. The first refers to protection, as in *se défendre contre une attaque* (defend oneself against an attack); the second refers to forbidding or prohibiting as in *défendre l'alcool* (forbid alcohol). The former is then subdivided into four parts, including the military meaning paraphrased by the verbs *protéger* (to protect, as in protecting a place) and *interdire* (to prevent, as in preventing via force or using weapons), again opposing protecting and forbidding. Le Larousse distinguishes five senses without hierarchization. One of the five is the legal use of *défendre* as in *défendre l'accusé devant la cour*, meaning to plead a case in court. Note that this sense was not included as a division in Le Robert. Disparity between dictionaries is not rare, nor specific to French. For an illustration of such a disparity in English, see the word *defend* in Webster's, The Oxford English Dictionary, and The Cambridge Dictionary.

Clearly, differences in how a word's meaning is separated into senses and how they are organized is indicative of the diversity of the trends found in lexicography in general, and also of the part played by the author(s). Lexicography—which is aimed at drawing up an inventory of the senses of words in context as well as over time—has been discarded as a linguistic approach to determining the conditions underlying language functioning. Thus, as obvious as it seems when one is reading a dictionary, the validity of the notion of polysemy itself has been subject to debate.

5.2.2 The Hypothesis of Polysemy

The question raised in linguistics to account for variations of meaning in context has gravitated around whether or not the semantics of a lexie has an internal structure. Fodor and subsequent authors [4] have defended an atomist point of view in which a lexical concept is an indivisible unit, based on the idea that lexical meaning is not defined by the word's usage context, nor by the potential productivity generated by that context, nor by the inferences one can attribute to it (*dog* $\longrightarrow$ *animal*) but by the lexical concept, i.e., its logical form, an atomic, mental unit. For example, "that which is *dog*" is whatever falls under the lexical concept *dog*. The possibility of describing the meaning of a lexical entry as something other than a list of homonyms is rejected, as is the possibility of finding what is in common and what is variable in the allegedly different meanings found in the various usage contexts. According to these authors, the main argument supporting this position is drawn from a confusion between the language level (that is, the usage level and context-variation level) and the denotative content in which a lexical unit's literal meaning is directly inscribed.

In addition to Fodor (but for different reasons), Rastier and Valette defended the idea that polysemy is an artifact [5]. This artifact is said to originate in a lexical approach, whereas semantics should be textual. Given that the construction of a sign (lexie) is itself considered to be artificial, the idea would not be to look at what follows from it, in terms of its stability and/or potential variations, but to make use of the dissemination of the text's semes so as to describe differences that are momentarily inscribed in (or between) various occurrences of the same lexie. The excerpt in the footnote below[1] describes this point of view more precisely [5].

However, for many authors like Pustejovsky [6] and also ones from the French continuist school [7], polysemy is a massive language phenomenon, as shown, moreover, by the complex structures found in the vast majority of dictionary and thesaurus entries. I address this point once again here, in the wake of the recognition of polysemy as a language-specific phenomenon. My reasons are as follows. The object of study herein is language, and more specifically lexical semantics, which does not

[1] "Polysemy appears [...] as a triple artifact: (i) that of a linguistics of the sign, which allows itself to consider isolated signs without raising the question of their discretization; (ii) that of an objectivist point of view, which presents the signifier as an obvious fact; (iii) that of a dualist metaphysics, which separates language and thought like matter and mind." (our translation).

deal with content but rather with the link between expression form and content. I contend that a lexie is a language unit that results from the connection between two heterogeneous spaces, the expression space and the content space, and that polysemy comes from the heterogeneity of these linked spaces. The notions of concept or even lexical concept, therefore, do not seem directly necessary (as we shall see later) or may even be misleading if they are substituted for units of meaning.

In the following sections, I will look at a few answers that have been forwarded in attempts to formalize lexical meaning and its variations.

5.3 Formalizations Proposed for the Study of Lexical Meaning

In lexical semantics, the various hypotheses and theories proposed delineate what a semantic unit must be and what formal unit it is assigned. Defining the "semantic unit" depends upon whether the linguist thinks that polysemy exists, and more specifically, whether the opposition between atomism and polysemy exists. As noted above, whenever atomism does not grant an internal structure to the semantics of a lexie, the alternatives consist either of drawing up a "frozen" list of meanings, each one becoming a semantic unit, or assigning each lexie a structure that can be used to compute its meaning in context. In the latter case, the meaning is not fixed but is constructed and adjusted based on a format that can contain both stable and variable elements.

Ultimately, one could consider, that the atomist approaches—by definition symbolic and discrete but not formalized to the limit—mathematize each lexical concept directly by a variable that fully represents it. In the approaches that assign an internal structure to lexical semantics, two aspects stand out. The first is representational: it defines the internal lexical structure of the lexie. The second is computational: it relies on the representation chosen to ground the computation of meaning in context.

5.3.1 Internal Lexical Structure

In certain proposals, the definition of the internal format of a lexie's semantics was constructed by fragmenting it into subparts of varying granularity. In others, the definition relied on a meta-language of features none of which, taken separately, was a sub-part of the lexical meaning.

Lexical Structures With Zero or Coarse-Grained Fragmentation

Although vector models, whether based on data analysis or deep learning, do not directly rely on the fragmentation of lexical meaning, I am mentioning them here because they serve as a basis for the computational models presented later. The representation unit of this paradigm is the vector, the unit of lexical meaning; it is defined by coordinates in a multidimensional space. In neural networks, for example, the unit of meaning is a state of the system (a weight vector corresponding to the linguistic entity given as input). Vector models thus implement an injection from the set of all language units in the entire constructed space. Although they are continuous, these approaches are mappings in the sense that each lexie (and its semantics) is associated with a unit of the model. There are no other semantic units constituting the lexical meaning than those defined by this projection. It follows that polysemy is not directly represented therein but results, *a posteriori*, from a computation.

Among the non-applicative approaches that strive to represent finer units of meaning than the lexie, we find course-grained finite approaches. As mentioned above, WordNet decomposes a lexie into a finite list of lexical concepts (Synsets), which are the units of the model. This approach, of logical inspiration, nevertheless remains based on hierarchical tree structures with a finite and fixed list of nodes. Each sense (Synset) and node of the tree is a unit that can be characterized and/or discriminated by its list of synonyms, and be accompanied by a gloss, a sort of definition expressed in natural language (as found in dictionaries). These units are positioned in an ontological tree supplemented by crosswise links representing types of relations other than hyper-hyponymy, namely, metonymy, antonymy, troponymy, and so on.

Lexical Structures With Fine-Grained Fragmentation

Among the fine-grained fragmentations of lexical meaning, whose constitutive semantic units cannot necessarily be characterized and discriminated from a definition or a gloss, we find geometric models, which are models of hypergraph[2] embeddings and context vectors of neural networks.

Geometric Modeling

Geometric models like the model used for Semantic Atlases (SAs) assume two levels of organization: that of lexies, and more specifically of their expression form, and that of their content, in which meaning divides the lexie's content into portions. I will look at this approach here because it supports what is developed in Part 3 of this book. Insofar as content is not directly accessible, it is approached in SAs by an artifact called cliques, which are minimal units of meaning. Cliques are complete, maximal sub-graphs of lexical links based on (para)synonymy or association.[3] Figure 5.1 illustrates the notion of clique as a sub-graph of a lexical relationship (here (para)synonymy) but also as the intersection of semantic domains (a sort of depiction

---

[2] Hypergraph $H = (V, E)$ refers to a pair where $V = \{v_1, v_2, ..., v_n\}$, a finite set of vertices, and $E = \{E_1, E_2, ..., E_m\}$, a family of non-null parts of V such as $\bigcup E_i = E$.

[3] Regarding association, which is based on regular co-occurrence and was renamed contextonymy, the reader may wish to refer to Ji and Ploux [8]. Co-occurrences are constructed via text scanning. Figure 5.3 shows the organization of the lexical meaning of the word *conductor* derived from contextonymy cliques.

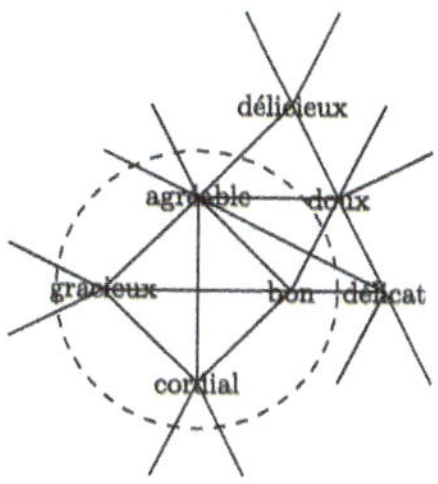

...
C_{100}: agréable, **bon**, cordial, gracieux
C_{101}: agréable, **bon**, doux, délectable, délicat, délicieux, exquis, savoureux, succulent
C_{102}: agréable, **bon**, doux, délectable, délicat, délicieux, exquis, suave
...
C_{270}: bien, **bon**, convenable, correct, honnête, honorable, satisfaisant, sortable
C_{2711}: bien, **bon**, convenable, digne, honnête, sérieux
C_{272}: bien, **bon**, convenable, honnête, honorable, moral
...

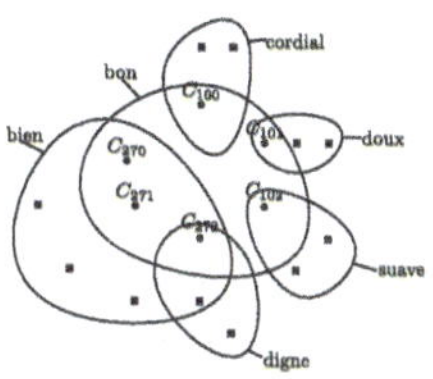

	bien	bon	cordial	...	doux	...	suave	...	digne	...
...										
C_{100}	0	1	1		0		0		0	...
C_{101}	0	1	0		1		0		0	...
C_{102}	0	1	0		0		1		0	...
...										
C_{270}	1	1	0		0		0		0	...
C_{271}	1	1	0		0		0		0	...
C_{272}	1	1	0		0		0		1	...
...										

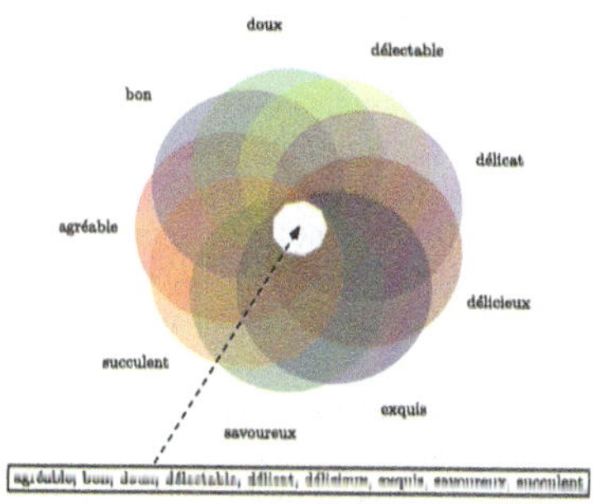

Fig. 5.1 Upper left: Part of the French synonymy network. Upper right: List of cliques extracted from all cliques containing the lexeme *bon* (good). Center left: Set of cliques taken as vertices of the hypergraph whose hyperedges are lexemes. Center right: Incidence matrix of the hypergraph used to compute a geometric embedding. Lower: A clique seen as the intersection of semantic domains of a list of synonyms

of a content form) associated to each synonym or parasynonym. Some examples of cliques in French are listed therein. A few properties of these units are given below.

Cliques are not language units. It is difficult to use a gloss to characterize the differences in meaning cover, as shown by the cliques in Fig. 5.1 and the following two cliques that contain the French word *maison* (house): (1) *abri* (shelter), *asile* (asylum, home), *habitation* (dwelling), *maison* (house), *toit* (roof), (2) *abri* (shelter), *habitation* (dwelling), *logis* (lodge), *maison* (house), *toit* (roof). Compared to the Synsets of WordNet, synonymy cliques are much more fine-grained. For example,

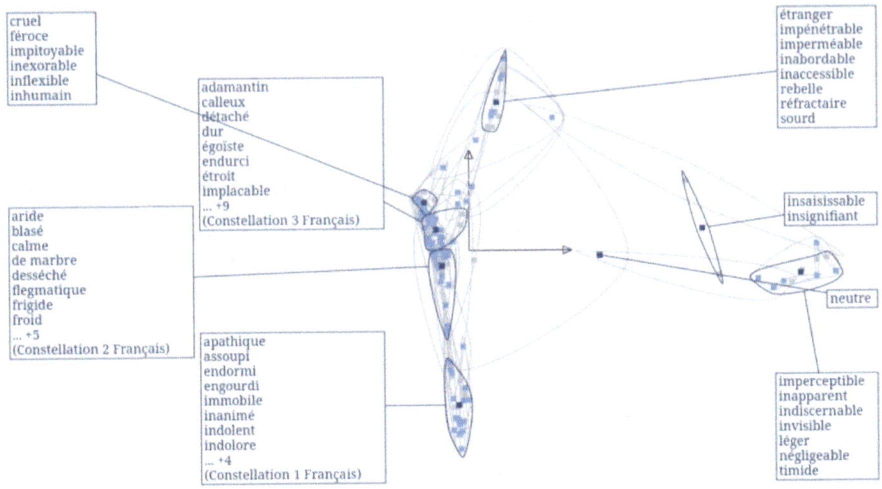

Fig. 5.2 Semantic space of the French word *insensible* constructed from (para)synonymy links. The gray outlines show the perimeters of the lexies, the points are cliques, and the blue outlines are classes computed by hierarchical classification with a Ward distance that distinguishes semantic values. On the right, we see the class containing the synonyms *imperceptible* (imperceptible), *inapparent* (unnoticeable), *indiscernable* (indistinguishable), etc., corresponding to the gloss "That which cannot be felt or perceived, or only barely so." (TLFI, our translation). On the left, we see the three classes computed here. The first (in the center) corresponds to the gloss "he or she who has no moral sensitivity, is incapable of emotion, sentiment, sympathy" (TLFI, our translation). It contains the synonyms *dur* (hard), *sans-coeur* (heartless), etc. The second (on top) corresponds to the gloss "Being insensitive to, remaining indifferent, not letting oneself be convinced. It contains the synonyms *réfractaire* (refractory), *étranger* (foreign), *sourd* (deaf), etc. The third (on the bottom) corresponds to the gloss "he or she who does not have (or no longer has) normal or usual physical sensitivity" (TLFI, our translation). It contains the synonyms *engourdi* (numb) and *indolore* (painless). Note that the geometry of this representation points out some clearly distinct values, like the value of *insensible* here (on the right side of the figure). This geometry also points out strong, potentially metaphorical proximities, as illustrated by the expression *un esprit engourdi* (a numb mind), which in certain contexts can be paraphrased as *un esprit insensible* (an insensitive mind), by bringing to bear the three values represented on the left side of the figure, and by mobilizing physical, moral, emotional, sentimental, and cognitive aspects

in English there are 2 Synsets for the word *insensitive* and 14 for the word *house*, whereas there are 9 and 82 cliques, respectively (for a more extensive comparison, see [9]).

Cliques are minimal units. The definition of a clique as a maximal sub-graph implies that there are no other lexies able to subdivide the intersection of the semantic domains of the synonyms that compose it. Synchronically speaking, then, these units are minimal.[4] The idea is thus to use this level of granularity by choosing the cliques as semantic units.

[4] Diachronically speaking, one can consider that since the content forms of lexies vary, their mutual intersections and thus their cliques vary too.

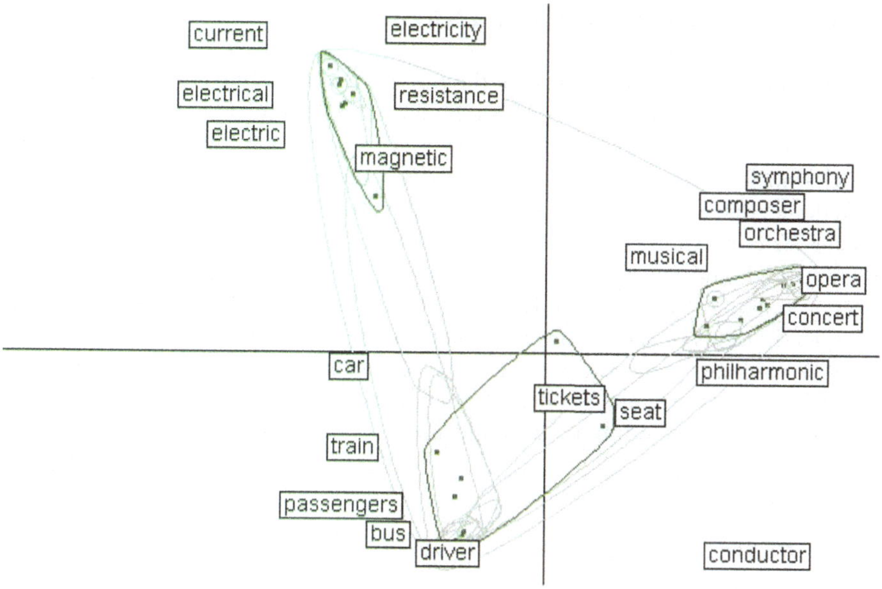

Fig. 5.3 Semantic space of the English word *conductor*, borrowed from Ji and Ploux [8] and constructed from cliques of contextonymy links, with embedding performed by a correspondence analysis. The gray "outlines" are the contours of lexies, the points are cliques, and the green outlines are classes computed by a hierarchical classification with a Ward distance. We can see three main classes here: that of a *conductor* in the sense of a train or bus driver, with the corresponding associated words; that of an orchestra conductor and the words associated with that meaning; and finally, in the upper left quadrant of the figure, the class related to the notion of electrical conductor. To illustrate how this representation interconnects semantic values, note that the words *tickets* and *seat* overlap with the classes related to transportation and the concert hall, and that they participate similarly in their respective semantic values

Cliques as content-form units. The semantics of a lexie is represented in the SA model by a geometric shape whose points are the cliques. The internal organization of the shape is computed by spatially projecting the cliques while preserving their mutual proximities. This projection was implemented as a factorial correspondence analysis (FCA) on a $clique \times lexie$ matrix but could be derived from other types of embeddings. I will return to this later. The resulting organization of the semantic shape points out the most generic semantic value, if one exists, and also the more specific values (see Fig. 5.2). Classifying the clique points makes it possible to find granularity classes that are closer to those found in dictionaries or WordNet (see Figs. 5.2 and 5.3), which are sometimes called semantic values and can be described and differentiated by a gloss. These topics are developed in greater detail in [10]. Note also that this approach applies to the entire lexicon, and in particular to adjectives, whose nature is unsuited to the ontological format of WordNet.

Context Vectors in Recurrent Neural Networks

After having noted that neural nets of the Bag-of-Words type like Word2Vec, which associate a vector to each word or lexie by formal mapping, do not allow us to represent its polysemy, various studies made use of recurrent (ELMO) or Transformer (BERT) models to compute vectors associated to contexts of use within a sentence. From these context vectors we can then—as for all of a lexie's cliques and in particular for the contextonymy clique—represent their mutual proximities or distances, classify them, and verify whether their grouping by neighborhood represents the different semantic values of the lexie. As recalled above, comparisons of meaning organized in terms of semantic value, and meaning organized in terms of context of use in the language, have often given rise to inconsistent results across studies. The idea, however, is the same: to obtain formal units (vectors) that represent particular uses, and by that token, fragments of meaning.

5.3.2 *Representation and Computation of Meaning in Context*

After showing that the phenomena of creativity of meaning in context cannot be studied on the sole basis of a fixed and finite list of semantic values, and having thereby underlined, for that same reason, the inability of dictionaries to account for that creativity, Pustejovsky proposed (in his Generative Lexicon) a generative model underlain by lambda-calculus, a logical framework. Each lexie is assigned a structure composed of a set of attributes and the possible values of those attributes. Operations control the semantic-composition or co-compositionality mechanisms, and update the values when the lexie is inserted in a phrase. Like generative grammar for the production of well-formed sentences, this approach—defined in a discrete formalism—takes its name from the desire to produce the set of allowable meanings by applying operations. Polysemy, here, is thus a computation, or more precisely a derivation, operating on a symbolic format in which the formal units directly represent the lexies. The lexical structure is constructed from a vocabulary of attributes capable of being interpreted from a linguistic standpoint, e.g., qualia, argument structure, etc.

Older neural networks [11] were also aimed at proposing a computational approach to polysemy. Meaning in context was defined according to a grid of parameters, here again, pre-set by the linguist. The network then learns to compute the values of the connection weights between neurons in such a way that the classification of polysemy it implements is consistent with pre-selected examples. However, in this case, the semantic values are continuous regions defined in the space of the chosen linguistic parameters and towards which the system converges when a usage context is entered for the lexie in question.

As recalled just above, the computation of polysemy in context is defined on the basis of the representational aspect chosen beforehand. In simple vector models, this computation is carried out by selecting or more generally weighting the word vectors. The meaning of the phrase or sentence in this case is itself computed as a vector [12], that is, as a linear combination of the word vectors that contain it. In non-probabilistic logical or neural paradigms, the definition of a formal structure of lexical meaning by metalanguage is used as a grid for adjusting the meaning in context. Each feature of the metalanguage is a linguistic property chosen by the linguist and will be attributed a symbol and possible values.

The preceding chapter pointed out two movements accompanying the mathematization of mechanics: the joint redefinition of the thematic and mathematical fields, and the separation of domain units and model units. The present chapter showed that the dominant models in linguistics and also in lexical semantics did not conform to the second movement of abstraction. The non-differentiation of linguistic units and model units that these movements involved necessarily induced limitations on any model proposed to represent and compute meaning. In particular, the same criticisms formulated above regarding auto-generative models also apply here. Additionally, we again find the same uniformity and homogeneity of models, and thus, a formal amalgam of lexical natures and functions. This amalgamation hinders the model's ability to distinguish them and deprives it of the ability to account for interactions between semantics and syntax.

Furthermore, the vast majority of models[5] separate paradigmatic relations from syntagmatic agreement by choosing to rely only on the former or the latter and thus depriving them of their necessary complementarity (see Sect. 2.2). In what follows and in line with the above considerations, I will attempt, to propose a new paradigm for modeling content forms that supports the phenomena of polysemy and composition. As stated above, this paradigm will be based on infralinguistic units and will combine paradigmatic relations and syntagmatic agreement. It will include antonyms, which have hardly been considered, if at all, in current models. Lastly, in the proposed model, specific operators will be assigned to the different parts of speech and to their syntactic functions.

References

1. Kamp, H., Reyle, U.: From Logic to Discourse (1993)
2. Lascarides, A., Asher, N.: Segmented Discourse Representation Theory: Dynamic Semantics with Discourse Structure. In: Computing Meaning, pp. 87–124. Springer (2007)
3. Leca-Tsiomis, M.: Les dictionnaires en Europe: présentation. Dix-huitième siècle, pp. 4–16 (2006)
4. Fodor, A., Lepore, E.: Remarks and Replies. The Emptiness of the Lexicon: Relections on james pustejovsky's the generative lexicon. Linguist. Inq. **29**(2), 269–288 (1998)

[5] Note that in addition to composition operations, Pustejovsky's proposal included lexical inheritance, which organizes lexical concepts into a network and thus includes paradigmatic relations. However, the implementation of the formalism he proposed remains problematic, for the same reasons as given on Sect. 1.1.1, namely, difficulty developing a logical system that we can be sure is both complete and consistent.

5. Rastier, F., Valette, M.: De la polysémie à la néosémie. Texto ! Textes et Cultures **14**(1), 97–116 (2009)
6. Pustejovsky, J.: The Generative Lexicon. MIT Press, Cambridge, Massachusetts (1998)
7. Fuchs, C., Victorri, B.: Continuity in Linguistic Semantics, vol. 19. John Benjamins Publishing (1994)
8. Ji, H., Ploux, S.: Automatic contexonym organizing model. In: Proceedings of the 25th Annual Conference of the Cognitive Science Society, pp. 622–627 (2003)
9. Ploux, S., Ji, H.: A model for Matching Semantic Maps between Languages (French/English, English/French). Comput. Linguist. **29**(2), 155–178 (2003)
10. Ploux, S., Boussidan, A., Ji, H.: The semantic Atlas: an interactive model of lexical representation. In: Proceedings of the 7th Conference on International Language Resources and Evaluation (LREC'10), Valletta, Malta, LREC (2010)
11. Victorri, B., Fuchs, C.: Polysémie et construction dynamique du sens. Hermès, Paris (1996)
12. Kintsch, W.: Predication. Cogn. Sci. **25**(2), 173–202 (2001)

Part III
The Dual Quasi-Fractal Tree-Like Structure in Lexical Semantics

I propose a new unit for modeling in lexical semantics: the dual quasifractal tree-like structure. This unit is chosen as a base to account for the notion of content form, its extent, and the linguistic phenomena related to it. In this paradigm, each unit has the potential to contain subparts, opposites, and a multi-scale organization. The term *tree-like structure* is preferred over that of *tree* because different subdivisions can be linked to the same sub-subdivisions. To improve readabilty in the rest of this book, the word *arborescence* may be used to refer to the concept of tree-like structure. The term quasi-fractal is preferred to fractal because the scales are finite. Finally, the term *dual* is used to group together all types of antagonisms, including opposites (*light* vs. *dark*) and contraries (*go* vs. *come*).

As an introduction, I will justify my choice of this unit format on the basis of analogies with the organization of living things and on the gain afforded by a multi-scale, tree-like organization. Then I will formally define the units using the notions of hypergraph and lattice, and I will show how they allow me to describe the link between heterogeneous milieus, here, between expression forms and content. The proposed tree construction method aligns with the structuralist approach in that it differentiates units on the basis of paradigmatic relations and syntagmatic agreement. However, the sets of differences and similarities in question here are systematized in such a way that the resulting units are no longer units of the language. The set of all dual quasifractal tree-like structures forms the semantic matter and replaces the vector spaces often used. Like the latter, this set must be equipped with operators to define the units' construction, modes of transformation and composition, and possibly their representation through embedding in homogeneous spaces. The following chapters detail some of these operators.

Chapter 6
The Dual Tree-Like Structure: Biological Grounding and Structural Gain

6.1 Tree-Like Structures, Living Things, and Language

The branching structure is an organic and functional mode that is highly present in nature and in the organization of living things [1, 2]. The dendrites and axons of neurons, the venous system, the bronchial tubes, trees and their roots, and streams or other types of water courses all have a tree-like structure in which each fork is accompanied by a narrowing of the ramifications or branches (using graph-theory terminology). A tree-like structure or arborescence is also a way of representing different abstract organizations. Some examples are the phylogenetic trees of the evolution of species, ontology trees of concept hierarchies defined by inclusion links, syntactic trees in the framework of generative grammar characterizing the scope of the governance, and the structure of a dictionary entry based on distinctions that go from the general to the specific.

In each case, a branch of the arborescence can split into lower-level branches, themselves associated with a spatial, temporal, or constitutive scale (or span) that is finer than the branch that subsumes it (in illustration, see Fig. 6.1, which applies the tree-representation rule established by Leonard da Vinci). Moving back up, we see that the lower-level branches converge into a branch with a larger temporal, spatial, and constitutive size. A key element in common is the distribution (or differentiation) vs summation (or integration) duality. The term *differentiation* refers here to the idea of being mutually exclusive or divided; that of *integration* refers to the idea of being incorporated or combined. This duality applies to both the form of the arborescence and the flow that is circulating within it. So the dendrites of a neuron, whose cross-sections get larger and larger as they converge toward the cell body (as in a system of rivers), and finer and finer as they move away, would be the grounds for integrating the excitatory or inhibitory electrical signals coming from the extremities. In the case of dendrites, the integration process is modulated by the distance travelled, with each of the signal's components losing power as it is propagated. In a dual fashion, the action potential output from the neuron would be propagated as a function of

S. Ploux, *Semantic Matter and Its States*, Lecture Notes in Morphogenesis,
https://doi.org/10.1007/978-3-032-22809-3_6

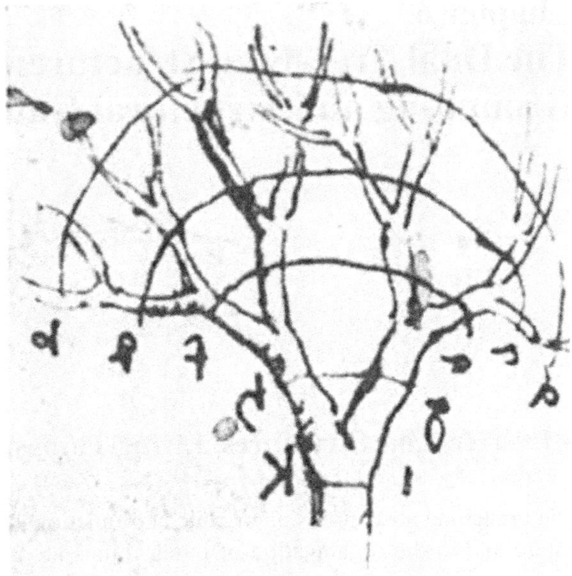

Fig. 6.1 Depiction of a mathematical equation governing the diameter of a tree's branches, according to Leonard da Vinci. "All the branches of a tree at every stage of its height when put together are equal in thickness to the trunk" (our translation). This equation has been corrected since then [4]

the tubular diameter along the axon and its ramifications. It has been shown, firstly, that the greater the axon's radius, the faster the signal is propagated, and secondly, that the morphology of the axon and dendrite tree structures is used to optimize information transfers and minimize their transmission time (see [3]). Analogous processes of integration/differentiation are assumed to operate throughout the human venous system and all along the branches and roots of trees.

When an arborescent structure is used as a means of representation but is not achieved materially, differentiation often looks much like specification. A subsumed element is a kind, a derivative of the subsuming element's potential. This is the case in conceptual ontologies and in phylogenetic or genetic trees. Recall that in conceptual ontologies, this general-to-specific or specific-to-general organization is echoed in the conceptual organization found in cases of impairment caused by localized brain damage (phenomena cited earlier [5]). The progression of impairments induced by semantic aphasia also supports this tree paradigm. It has been shown, for example, that patients with the semantic variant of primary progressive aphasia (PPA) tend to over-generalize: particular concepts and their specific attributes (e.g., the stripes on zebras) are lost first and replaced by more generic or prototypical ones (horse for any equine). This impairment, which evolves in a more marked way for categories of living things than for non-living ones, gradually affects increasingly broader categories (see for example [6, 7]).

I propose to show that a model based on the notion of a quasi-fractal arborescent structure in which differentiation is based on variable scales is a more suitable framework for lexical semantics, all the while taking up less space and using less computation than a model whose units are embedded in a space that is totally homogeneous, as are vector spaces. To do so, I will begin by bringing to bear another domain, that of signal processing, to recall the benefits of multi-scale processing.

6.2 Parsimony of Arborescent Structures as Opposed to Vector Spaces

6.2.1 The Example of Image and Signal Processing

Multi-scale wavelet analysis has enabled the compression, reconstruction, and analysis of signals such as images and sounds. Ingrid Daubechies (whose general-public lectures are always fascinating yet easy-to-understand) showed how an image, that is, a matrix of bounded values (ranging from 0 to 255 for a grayscale image), can be associated with an arborescent structure via a bottom-up process involving two operations on neighboring pixel values: averaging (or convolution with a wavelet of sum one) and difference computation (or convolution with a wavelet of sum zero). At each node, the arborescence contains the result of one of the two operations applied to the subsuming node: this operation reconstructs the initial image using a top-down process which, starting from the terminal node, integrates by summation, the various submatrices associated with each of the arborescence's nodes. This widely studied process, which has been generalized to include different types of wavelets, offers the advantage of compressing the original data, thereby reducing storage requirements. The compression process is based on the observation that images contain far fewer contrasts than regularities. Consequently, most of the differences computed at each stage are close to zero, allowing some to be disregarded without substantially impacting the reconstruction quality.

Furthermore, beyond the benefits of compression, the coefficients (or values) computed at each stage of averaging and difference computation can be used to analyze the image's characteristics by separating details from larger patterns in a multi-scale arborescence, locating contours or singular points, and so on. The move from a vector-based (or matrix-based) homogeneous mode, like that used for digital images, to a multi-scale arborescent mode thus offers numerous possibilities.

6.2.2 Multi-scale Processing: The Case of Vision

It has been noted that convolution networks are inspired by and share signal processing analogies with those involved in vision [8]. Indeed, V1, the first area of retinal projection onto the cortex, performs local, retinotopic processing that is then propagated from area to area, each area supporting increasingly global and complex and less localized processing. The comparison between image processing and vision nevertheless has its limitations. Clearly, unlike a digital or printed image, which forms a uniform matrix of values, the organization of the initial retinal stage has a concentric neuronal density around the fovea and macula, where detail detection takes place. Thus, a flat and uniform image is not input directly into the visual neuron system. Despite the difference in input format, neuronal processing progresses from details to more global patterns. Note that in compression tree structures like those briefly described above, the patterns associated to wavelets are neither delocalized nor complex in terms of visual processing. However, other systems known as pattern-recognition systems reproduce non-localized processing analogous to that of vision. These systems are aimed not only at encoding and reconstructing but also at classifying (or recognizing) a set of images through example-based learning. Their architecture includes a series of neural layers that perform convolutions of an increasingly large span (see for example [9, 10]).

Just as the format of input to the visual cortex does not have the uniformity of a matrix-like image, if the processing units of vision have spans of variable complexity, then a uniform matrix-like format is not a specifically neural representation. If people have the illusion of seeing things in a uniform space, then we need to provide an explanation for that illusion. We could explain it terms of a top-down projection onto one and the same surface, and potential multi-scale processing going from the most global and complex to the simplest and most local. The projection surface could be the retina or an artifactual surface (photograph, sheet of drawing paper).

The example of multi-scale image and vision processing serves here as a guide and illustration of the advantages it provides. Below, I will attempt to construct an arborescent structure of the content form that will allow me to point out its benefits over a vector-based representation in a totally homogeneous space.

6.3 Content Form and Dual Quasi-Fractal Arborescent Structures

The content form links two heterogeneous milieus. Note that this is also true in the domain of pattern recognition, which associates a pixelated image to a known category, such as those of different animals, plants, etc. However, in artificial neuronal networks that do pattern recognition, categories are represented by probability distributions and images are represented by intensity matrices. This amounts to forcing homogenization of the different milieus (one that encompasses categories, the other that encompasses pixelated images) in real ($\mathbb{R}^n$) or complex ($\mathbb{C}^n$) spaces of the same nature, from which it becomes possible to perform computations that include both input and output. Maintaining my idea that expression forms and content are heterogeneous entities prevents us from constructing units with a shared nature and of conceiving of the link between them as a function defined within one and the same encompassing space. This amounts to starting from a source milieu (here, the expression forms) without access to the materiality of the target (its content) and performing operations of variation or difference on the source units, and then assessing the contrast vs stability effects that emerge in the target milieu. In line with structuralists, then, I will start from operations on expression forms and the confirmed or unconfirmed effects of those operations on content,[1] in order to construct the corresponding content forms.

Hooking up expression form and content, and ramifications of the content form. It is hypothesized here that a content form is constructed from initial hook-ups or connections made between an expression form and a content regularity, and then evolves through a series of differentiations or integrations (in the sense stated above). These differentiations and integrations themselves depend on cognitive processes and internal or external contingencies. For illustrative purposes, take the example of a little girl who, having heard or said repeated associations with content, co-constructs a sign made up of an expression form and a content form, and thus learns a few initial words. Let's choose the word *dog*. After making the connection between elements from heterogeneous milieus, the links between these two facets of the sign (a) will become enriched (via the integration and generalization of the perceptions associated with *dog* (the word and the animal), (b) will be narrowed down due to competition with the signs of other canines like wolves, (c) will branch out (a live dog vs other possible categorizations like a stuffed or toy dog), and (d) will expand (*He's a dog!* to insult someone). These transformative adjustments are characteristic of the creation of links between heterogeneous milieus, wherein, and between which, the morphisms are precisely tenuous. They depend on recognition of contrasts or stability. As the lexicon expands each content form can cover a highly distinct content (*to eat* vs *a cloud*), one that is partially shared (*animal* vs *dog*) or nearly identical (*creature* vs *animal*).

Antagonisms. In Wallon's The Origins of Thought in the Child [11], the author noted that "every expression, every notion is intimately united with its opposite, in such a way that it cannot be thought without it. It is from its opposite that it finds its *raison d'être*, that is, its primary meaning, its most elementary, its coarsest but most essential specification..." (p. 67, our translation). Indeed, very early on, the child produces refusals expressed by *No*. Also very early on, the child acquires contraries (*big* vs *little*) and opposites or complementaries (*come* vs *go*, *outside* vs *inside*). Here then, we consider opposites to be consubstantial with the semantics of a lexie. So a lexie's quasi-fractal arborescence will necessarily integrate them in a dual form when ever they exist.

True contrasts and benefits. A content form is constructed on the basis of contrast effects or stability effects that impact the content or the expression form. The contrasts hinge on general cognitive processes such as the subject's expectations,[2]

[1] Conversely, if measures of brain activity were fine enough (which current imaging methods approach without yet achieving the granularity of measures of language effects like polysemy), we could consider starting from the neurophysiological signals of contents and their locations, recorded during speech, in order to correlate their disparities with those of the expression forms emitted.

categories, which for some, precede experience (as in the opposition *living* vs *non-living*), lexicon enrichment, and competition between words, etc. They also depend on more specific contingencies related to each person's particular history. As in image compression, these contrasts are true contrasts in the sense that they are based solely on contents that are partially shared, are opposing, or are related by paradigmatic or syntagmatic relations. The idea I am defending here is thus that the content form of a lexie, sentence, or text (i.e., its meaning) is not described within a single, complete system with a fixed number of parameters—which, by that token, would necessarily be very large—but within a system of semantic proximity that disregards irrelevant contrasts. As with images, this approach makes the representation space smaller.

A content form and its parts are not language units. As recalled above, in this construction and evolution of a content form, the ramifications are not language units. They are semantic entities of a finer grain than the entire content form associated with an expression form (lexie), and may be located below any possible naming level. In this sense, they may not be directly accessible or characterized via a definition or a gloss expressed in the language. The upcoming chapters are devoted to the implementation of the notion of a content form's arborescence introduced here.

References

1. Ragan, M.A.: Trees and networks before and after Darwin. Biol. Direct **4**, 1–38 (2009)
2. Mandelbrot, B.B.: The Fractal Geometry of Nature, vol. 1. WH freeman New York (1982)
3. Desai-Chowdhry, P., Brummer, A.B., Savage, V.M.: How axon and dendrite branching are guided by time, energy, and spatial constraints. Sci. Rep. **12**(1), 20810 (2022)
4. SBD Sopp and Ruben Valbuena: Vascular optimality dictates plant morphology away from Leonardo's rule. Proc. Natl. Acad. Sci. **120**(39), e2215047120 (2023)
5. Warrington, E.K., Shallice, T.: Category specific semantic impairments. Brain **107**(3), 829–853 (1984)
6. Laisney, M., Giffard, B., Belliard, S., de La Sayette, V., Desgranges, B., Eustache, F.: When the zebra loses its stripes: semantic priming in early Alzheimer's disease and semantic dementia. Cortex **47**(1), 35–46 (2011)
7. Mesulam, M., Rogalski, E., Wieneke, C., Cobia, D., Rademaker, A., Thompson, C., Weintraub, S.: Neurology of anomia in the semantic variant of primary progressive aphasia. Brain **132**(9), 2553–2565 (2009)
8. LeCun, Y., Bengio, Y., Hinton, G.: Deep learning. Nature **521**(7553), 436–444 (2015)
9. Mallat, S.: Understanding deep convolutional networks. Philos. Trans. R. Soc. A Math. Phys. Eng. Sci. **374**(2065), 20150203 (2016)
10. Taylor, G.W., Fergus, R., LeCun, Y., Bregler, C.: Convolutional learning of Spatio-temporal features. In: 11th European Conference on Computer Vision. ECCV 2010, pp. 140–153. Springer, Berlin (2010)
11. Wallon, H.: Les origines de la pensée chez l'enfant. Les Etudes Philosophiques **2**(1) (1947)
12. Thom, R.: Esquisse d'une sémiophysique: Physique aristotélicienne et théorie des catastrophes. InterEditions (1991)

[2] Thom [12] spoke of *pregnance*, a term he used to refer to forms that have biological significance.

Chapter 7
Computation of Content-Form Arborescences

In this chapter, I propose a way of computing and constructing the arborescent structures associated with the semantics of a word or a set of units of meaning. These structures are built from paradigmatic relations, to which I add antonymy. From a mathematical standpoint, the arborescences are hypergraphs. Their construction also endows them with a lattice structure. After defining them formally, I will discuss the possibility of embedding the tree-like structures in geometric or topological spaces, with the aim of preserving the multi-scale structure induced by the lattice. However, I will emphasize the generic nature of the arborescent format in light of the multiplicity of possible embeddings. Syntagmatic arborescences and the operators used to compute lexical composition are the topics of Chap. 9.

7.1 Content Forms and Paradigmatic Relations

As previously noted, content forms are tied to a set of heterogeneous content items, insofar as they connect expression-form milieus to content milieus. Their heterogeneity requires a directional approach based on measures of the variation effects of one milieu on the other. Like the structuralists, my approach involves starting from expression forms, and then processing them in order to assess their effects on content, without representing that content.

Here, I call upon two ways of processing (hereafter called *manipulating*) expression forms, to detect content sharing: substitution and permutation, both of which are presented in Part 1 of this book. I recalled therein that assessing the stability of meaning by substituting one lexie for another in sentence or phrase contexts is a way of detecting synonymy or parasynonymy relations. Similarly, determining whether

S. Ploux, *Semantic Matter and Its States*, Lecture Notes in Morphogenesis,
https://doi.org/10.1007/978-3-032-22809-3_7

permutation is possible (or impossible) in sentences such as *a dog is an animal vs* *[1]*an animal is a dog* or in ones that contain a lexical anaphor like *my neighbor adopted a big dog; That animal is not very friendly vs * My neighbor adopted a big animal; that dog is not very friendly* enables detection of hyper- or hyponymy relations. To synonymy and hyper- or hyponymy, let me add antonymy, which is sometimes likened to a paradigmatic relation. Recognition of such relations is achieved by proposing often-short contexts in which substitution points to an opposing meaning (e.g., an *immense will vs a weak will*). Other meaning-reversal effects exist, such as those occurring in negative forms of sentences. This last point will be addressed in Chap. 8 on operators.

Extracting paradigmatic relations (synonymy, parasynonymy, hyper- or hyponymy) and antonymy are the result of the aforementioned manipulations and estimations of the stability or—on the contrary, differences—in meaning of the concerned sentences or phrases being manipulated. Without directly reproducing this linguistic analysis, I use some existing lexical databases of synonymy, hyperonymy, and antonymy to construct the paradigmatic arborescence representing a lexie's content form. As mentioned in Sect. 5.2.1, lexicographers' estimations vary. So the synonym database chosen here was compiled from several dictionaries, which allows for the integration of that variability. The synonymy relations are *symmetrized* therein. The symmetrization process is based on the observation that if one lexie is synonymous with another, then the reverse is also true, since there exist usages in which substitutions of one for the other preserve the overall meaning. The symmetrization process is also consistent with the idea that sharing between content forms is an intersection operation (Fig. 1.2), which itself is commutative. For hyper- and hyponymy, I rely on the direct hyperonymy relations (i.e., ones without intermediate nodes in the ontological tree) given by thesauruses in WordNet format [1], and more specifically for French, on the Wolf database [2]. Lastly, for antonyms, I use a custom-built database, which is also symmetrized. The makeup of this database rests, in part, on the automatic detection of antonymy candidates extracted from large literary corpora and newspaper articles, followed by verification of their relevance.

7.1.1 Content, Categories, and Paradigmatic Relations

Before continuing with how content forms are constructed using paradigmatic relations, let me first stress that synonymy relations are fewer in number and less refined for lexies referring to concrete things or animal species, and that hyper- and hyponymy relations are mainly relations between nouns, especially nouns referring to living things. Thus, except for a few pro-verbs like *do*, *be*, and *have*, the set of all verbs is not organized in a hierarchical way but rather according to networks representing partial semantic overlapping. This is also the case for adjectives and adverbs.

---

[1] An asterisk in front of a sentence means that it is semantically incoherent.

Lastly, I remind the reader that because opposites are mostly based on a dynamic of reversibility (e.g., *ascend vs descend*) or, for adjectives, on an order of intensity or degree (*big* vs *small*), it follows that lexies referring to concrete things also have few antonyms. Words associated with concrete things are better grasped via relations of syntagmatic agreement. Accordingly, the lexeme *pen* will be endowed with a paucity of paradigmatic relations but a wealth of syntagmatic agreement relations that refer to its functional, constitutive, or perceptual characteristics.

In what follows, I will describe how paradigmatic and syntagmatic tree-like structures work together to address the entire set of cases distinguished here.

7.2 Construction of Paradigmatic Arborescences

Once the synonyms, parasynonyms, and direct hypernyms have been set, the terminal elements of the tree structure are computed. These elements are seen as minimal-span anchors to a content item. Here the term "minimal" refers to the idea that through the systematic differentiation and sharing of content forms, no other lexie than those already involved can—for a given state of the language—separate content that has already been individualized. As in the geometric model of the Semantic Atlases, there are two equivalent ways of presenting these terminal elements. The first way relies on hypergraphs of paradigmatic relations. In this case, a terminal element of the arborescence will be a maximal clique of the paradigmatic relations and a vertex of the hypergraph. In the second (geometric) way, a terminal element is a non-empty, indivisible intersection of contents shared by a set of lexies, as depicted in Fig. 5.1.

The first step consists of computing the full set V_c of terminal elements and the vertex of the hypergraph. In the second step, hyperedges are added. For this, the notion of lattice is used. A lattice is a set E equipped with a partial or total order in such a way that for every pair of elements a and b in E, there exists an upper bound and a lower bound for the set $\{a, b\}$. The upper bound, denoted $\sup(a, b)$, and the lower bound, denoted $inf(a, b)$, can be described as two internal laws $a \vee b = \sup(a, b)$ and $a \wedge b = \inf(a, b)$. These two laws have the following properties: (a) they are commutative and associative; (b) for all a and b in E: $a \wedge (a \vee b) = a = a \vee (a \wedge b)$. The lattice constructed here has the inclusion relation as its partial order. It is defined as the smallest lattice, under the laws of intersection and union operations, that contains the maximal cliques. Every element of this lattice, except for the maximal cliques, is a hyperedge of the hypergraph. Note here that intersection and union are the inverse of each other, whether the element is represented in terms of cliques (maximal or not) or in terms of the geometric domain associated to them. For example, cliques $\{m_1, m_2, m_3, m_4\}$ and $\{m_1, m_2, m_3, m_5\}$, composed of lexemes m_i, have, as their intersection and subsuming node in the lattice, the element $\{m_1, m_2, m_3\}$ whose geometric domain completely covers the union of their respective geometric domains. Built as such, the hypergraph is a formalization of the arborescent content form. Figure 7.1 outlines the steps for constructing the hypergraph of the French word *acharnement* (relentlessness). It should be noted here that for a hypergraph to be a genuine lattice, one must assume the existence (unless there already is one) of a higher node $\top$ that can therefore be interpreted as representing the abstract notion of "lexie" and a lower node $\perp$ that can be interpreted as representing a unit of content that cannot be discerned from the union and intersection operations and thus has a span smaller than that of a clique.

In this method, one can find analogies with Fitch's [3] algorithm, which he used to draw up phylogenetic trees, and more generally, the trees found in the biology of evolution. However, in the model I propose here, we are not dealing strictly with a tree structure because a node can have several direct subsuming nodes.

7.2.1 Dual Arborescence

In order to maintain the necessary tension between a meaning and its opposite, each content form is associated with an opposing form. This opposing form is based on antonymy links organized into an arborescence based on paradigmatic relations. I will briefly present an illustrative example of such a construction. Take the word *insensitive*, which has many antonyms, including (in alphabetic order): *caring*, *charitable*, *compassionate*, *gentle*, *humane*, *kind*, *kindly*, *merciful*, *responsive*, *sensitive*, *sweet*, *sympathetic*, *tender*, *warm*. The idea here is to construct an arborescence that links their content forms as antagonists of *insensitive*. Accordingly, *sweet* is an opposite of *insensitive* when its semantic value refers to a being, usually a person. In this case, one could find usage contexts in which it would be possible for the antonym of *sweet*, to substitute some of its synonyms, such as *caring* or *humane* without changing its meaning. So the task is to select the part of the content form of the word *sweet* that opposes that of *insensitive* and not the part that *sweet* shares with *sugary*.

A criterion is defined (Fig. 7.2) in order to select from among the set of maximal cliques formed from antonyms in a paradigmatic arborescence and from their "similars"—based on synonymy, parasynonymy, and direct hyper- or hyponymy—from among those that are opposites of the units of meaning contained in this same arborescence. This criterion is general. It can be used, as we shall see later, for lexical relations other than antonymy. After the selection, the opposite arborescence is generated using the smallest-lattice algorithm presented above. The result obtained for the arborescence opposed to *acharnement* is given in Fig. 7.3. The one obtained for the French equivalent of the lexeme *insensitive* (insensible in French) is too large to print in this book.

7.3 Some Foundational Properties of the Model

Here, I will return to some properties of the model in light of the objections raised in Part I and Part II of this book.

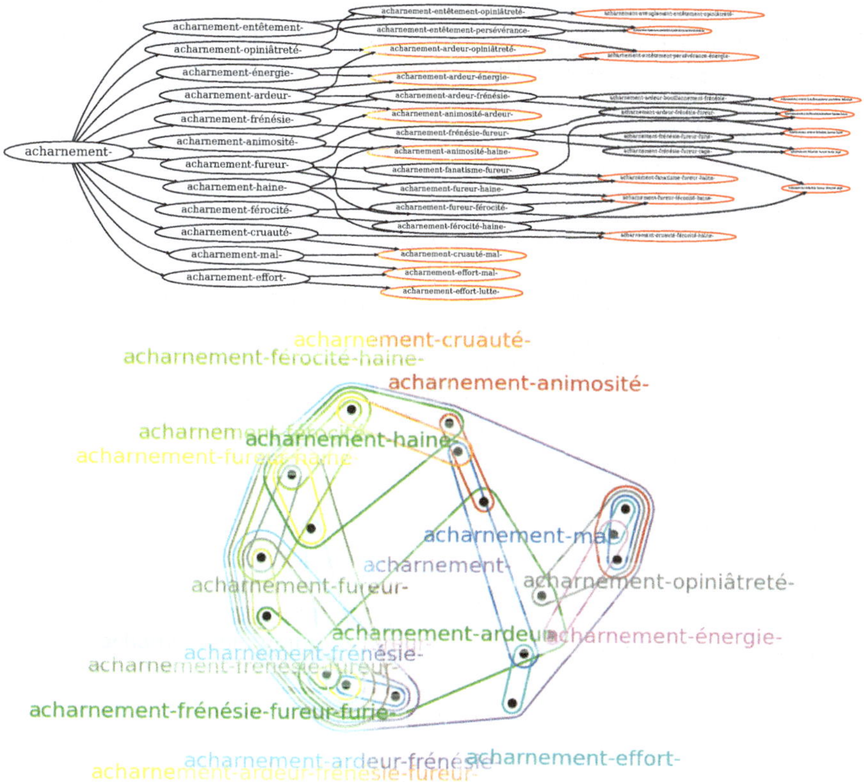

Fig. 7.1 (Previous page.) Upper: Paradigmatic arborescence of the French word *acharnement* (relentlessness), hereafter denoted $A_{acharnement^{p}}$, created by a hypergraph organized as a lattice. The maximal cliques are the leaves of the structure (in red). The other nodes of the arborescence are hyperedges of the hypergraph. The arborescence is constructed by calculating the smallest lattice (according to the intersection and union laws) that contains the maximal cliques. It begins by taking a bottom-up route, starting with the maximal cliques and calculating their intersections two by two. The maximum-degree vertices, $d_{\max}$ (here, the degree corresponds to the number of vertex or hyperedge elements) are then connected to the intersections that contain them and have $d_{\max} - 1$ elements. This process is reiterated, starting with the vertices or intersections of degree $d_{\max} - i$, for i ranging from 1 to $d_{\max}$. The missing links that ensure closure by intersection and union, a defining property of a lattice, are then added. Lower: The same hypergraph in geometric format (obtained by using the Python HyperNetX library), where the domains of each hyperedge are represented. Note that inclusion in the lattice-type of format (upper) is reversed in the geometric format (lower). Lastly, the polysemy of the word acharnement can be represented here. For example, the meaning of acharnement that shares the notions of perseverance or obstinacy, as in "Il faut travailler avec acharnement" (One must work relentlessly) or the meaning that shares the notions of rage or fury, as in "Tant d'injuste acharnement me pousse au désespoir" ("So much unfair relentlessness pushes me to despair", [4]) are distributed across the upper part of the arborescence for the former nuances of meaning, and in the lower part, for the latter

Let $\mathcal{R}$ denote a relation between language units. A matching criterion is chosen to ensure a good distribution of $\mathcal{R}$ between two cliques, c_i and c_j. The criterion used here is as follows.

Let M_{c_i,c_j} be an $n_i \times n_j$ matrix where n_i is the size of clique c_i and n_j the size of clique c_j. M_{c_i,c_j} contains 1 in row k and column l if $\mathcal{R}(c_i^k, c_j^l)$ for c_i^k in c_i and c_j^l in c_j; if not, it contains 0. And let $si_{1..n_i}$ (resp. $sj_{1..n_j}$) be the sums of the rows and columns, respectively, of M_{c_i,c_j}. Then the two cliques c_i and c_j are matched if $rank(^{M}_{c_i,c_j}) > min(r, max(n_i, n_j))$ with $r = 3$ or if $si_{1..n_i}$ (resp. $sj_{1..n_j}$) is $> \nu$ for at least $\mu * n_i$ rows (resp. $\mu * n_j$ columns), with $\nu = 1$ and $\mu = 2/3$, $\nu = 2$ and $\mu = 1/3$ and $\nu = 3$ et $\mu = 1/6$. The threshold values given here are used for the examples in the book.

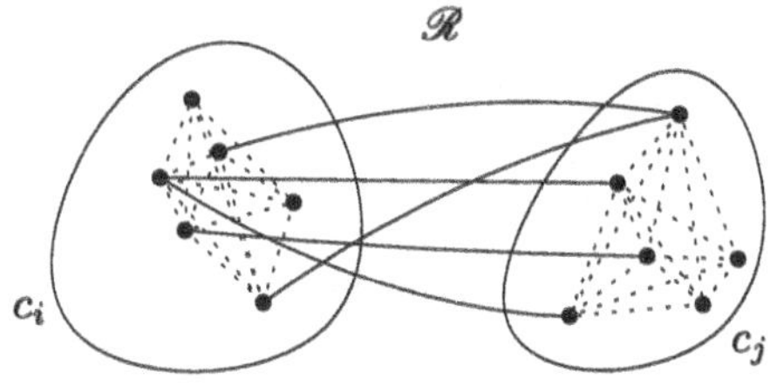

By applying this criterion with the antonymy relation, the following two cliques are matched (and therefore have opposite meaning here):

c_i: *acharnement, ardeur, bouillonnement, excitation, frénésie* (relentlessness, ardor, bubbling, excitement, frenzy)

c_j: *calme, flegme, patience, tranquillité* (calm, phlegm (in the sense of calmness of temperament), patience, tranquility)

The following two cliques are not matched:

c_i: *acharnement, ardeur, bouillonnement, excitation, frénésie* (relentlessness, ardor, bubbling, excitement, frenzy)

c_j: *affection, douceur, tendresse* (affection, gentleness, tenderness)

Fig. 7.2 Definition of the method used to match cliques, the terminal elements of two arborescences

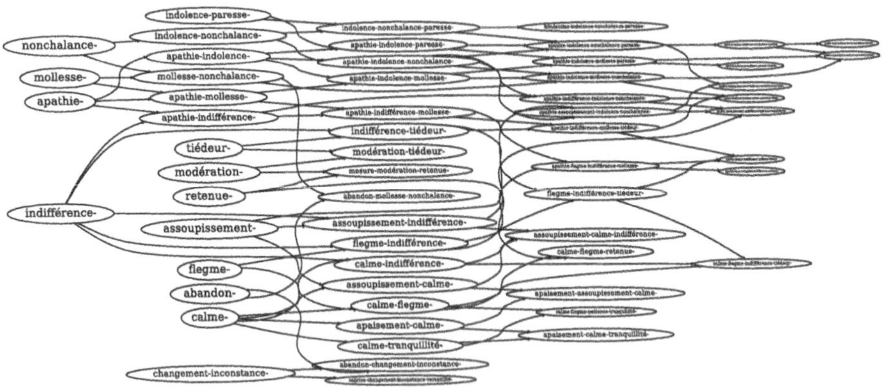

Fig. 7.3 Opposite arborescence of the French word *acharnement* (relentlessness), hereafter denoted $A_{acharnement^n}$. *Indifférence* (indifference), *apathie* (apathy), and *nonchalance* (nonchalance) are the highest-order hyperedges (Ord)

7.3.1 Representing Polysemy

The content form of a lexie is represented here as a complex arborescence that organizes the entire set of that lexie's meanings, with major branches at the top and increasingly finer distinctions descending from them. Thus, the number of vertices (nodes) or terminal elements that can be reached from a given hyperedge e, denoted here as $Ord(e)$, is an indicator of its polysemy. Figures 7.1 and 7.3 illustrate this aspect. In fine, the more generic semantic values are those represented by hyperedges with the greatest Ord and are closer to a root of the tree, while the more specific ones are represented by hyperedges of the smallest Ord values and are near the vertices or terminal elements.

7.3.2 Porosity of Meaning

One of the difficulties in representing the meaning of a lexeme arises from the porosity of its senses. In dictionaries, the structure consisting of finer and finer distinctions separated each time runs up against the issue of potential semantic overlap between the resulting divisions. An arborescent structure, in contrast, allows for representation of shared distinctions or branches that might join each other at a lower level.

7.3.3 Dual Content Forms

The semantics of a lexie are also characterized by it opposites. The dual content form, made up of both the manifested content form and its opposite, provides support for this relation, which—as recalled earlier—is essential and constitutive of the development of language in children.

7.3.4 Granularity of Meaning and Non-Linguistic Entities

Each level has a span with a finer granularity than the subsuming level. As stated in Chap. 5, the granularity of terminal and intermediate levels does not necessarily allow for characterizing and glossing their linguistic meaning. The model's entities are thus based on the construction of infra-linguistic units.

7.3.5 *A Single Global Space* vs *Local Spaces*

In vector models, units have only one format. The space is complete. Any word or any sentence can be situated with respect to any other. Here, the model is built on proximities or differences between units linked by a relation. It is therefore associated with the portion of the lexicon being studied. I will come back to this property in the conclusion, where it will be considered in terms of its cost in number of coordinates and parameters in language models.

7.3.6 *Genericity of the Model*

To conclude this list, and before addressing their embedding, I would like to emphasize the fact that the arborescent paradigm presented here is independent of the spatialization mode used (and for that reason, is more generic), all the while allowing for a representation of lexical meaning, and as we shall see, its computation in context.

7.4 Searching for Geometric Embeddings

Embeddings in $\mathbb{R}^n$ spaces are widely used in computational linguistics and even constitute the foundation of vector models. As previously mentioned, such embeddings enable not only computations of proximity that can be likened to similarity, but also a classification of embedded entities, among other things. When the dimension of the projection space is low, they offer the advantage of providing a concise view of the main characteristics of the relations between entities. These same reasons remain relevant in the multi-scale model of content forms. However, in this case, the embeddings are geometric and require preserving the order in which the lattice elements are included. Thus, the domain (area in the projected space) associated with a node of the tree will contain the domains associated with the nodes it subsumes. Similarly, when two nodes share subsumed nodes, this will imply the overlapping of their respective embedding domains. There is a formal condition for preserving the lattice structure during geometric embedding, namely, the lines connecting the centers of the areas associated with hyperedges must not cross each other. This condition is not always satisfied in the embedding methods used in the literature. A case in point is the embedding carried out by the Semantic Atlas model, which is built solely from lexical-relation cliques. Here, I will highlight the implications of this limitation and suggest some alternative directions.

7.4.1 Limitations of Extending the Geometric Model of Semantic Atlases (SAs)

Previously, in the geometric model of SAs, a factorial correspondence analysis (FCA) was applied to the incidence matrix of the hypergraph whose vertices were maximal cliques and whose hyperedges were lexies, thereby allowing for clique embedding. It is thus possible to use the same method to embed the terminal elements of a paradigmatic arborescence. The opposite hypergraph can be embedded in a space of its own, and the content form's leaves can be matched with the opposite-form's leaves, as shown in Fig. 7.2. The reduction of the dimension of the embedding space on the initial FCA axes is computed so as to best preserve the inertia of the cloud formed by the maximal clique points. In this case, starting from the arborescence's leaves, each node can be represented by the envelope of the embeddings associated with the nodes it subsumes. The root lexie is the overarching domain within which each subdivision of the arborescence outlines a subdomain.

However, reducing the dimension of the space by selecting the principal FCA axes can generate irrelevant overlaps of the arborescence-node embeddings. For instance, a maximal clique (or minimal unit of meaning) might have inadequate embedding on the initial axes. This could be due to the fact that it is associated with a highly specific meaning, in which case it would appear in a marginal part of the arborescence and on a high-dimensional axis of the FCA. Lacking any significant coordinates on the principal axes, it would show up at their center, which, if it exists, is the location of the most generic semantic value. It follows that the domain associated with this unit could be included within, or partially superimposed upon, the domain of nodes of the arborescence that do not subsume it. Stated differently, the projections of the lattice branches linking the domains associated with the hyperedges would cross.

Moreover, if, in an attempt to avoid space-reduction effects, we keep the full unreduced initial space, the resulting embedding would take place in a high-dimension domain whose dimension would be equal to whichever is smaller, the number of maximal cliques or the number of lexemes, and it would also be very sparse. In this case, the synthetic and visual advantages that the embedding was intended to afford would be lost.

Another limitation stems from method's ineffectiveness when applied to trees that can be partitioned into disjoint sub-trees. This is true, for instance, in cases of homonymy. A French lexie like *page* has several totally distinct values. These values are depicted by sub-trees which themselves are disjoint (see Fig. 7.4). In this case, when the FCA is performed, the embeddings of the homonym sub-trees end up being "stuffed" inside themselves. Also, the representation of the embedding is difficult to read and does not exhibit the desired spatialization qualities.

Several routes could be explored to improve the choice of embeddings described above. In the FCA, distances other than chi-squared could be used,[2] as well as non-

[2] It is possible, for example, to equip each arborescence with a distance computed from the shortest path linking two nodes. Here, in order to retain the scales of the arborescence, it would be consistent to weight the edges in accordance with the scales.

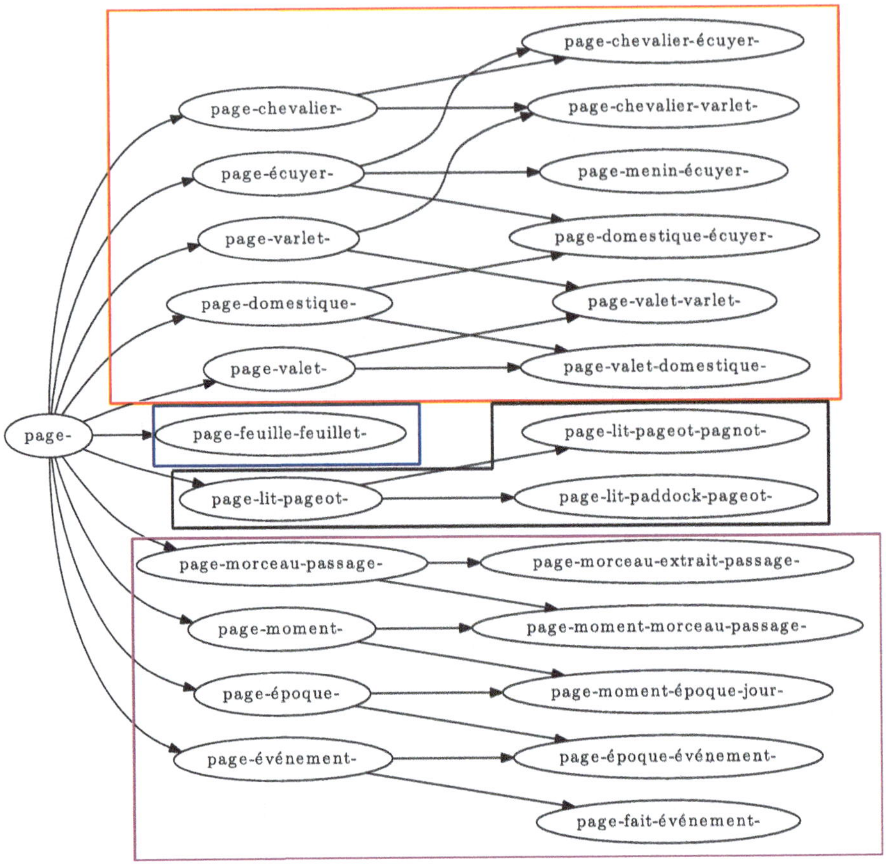

Fig. 7.4 The arborescence A_{page^p}, of the French word *page*, constructed from paradigmatic relations. The colored boxes separate the partitions of the hypergraph once the root has been removed. The partitions correspond to different homonymic values of the word, from which prototypical representatives can be extracted: *le feuillet* (the leaflet), *le valet* (the jack), *le moment* (the time), *le lit* (the bed, a regional use)

linear space-reduction alternatives like $t - SNE$ [5] or UMAP [6]. I will not give a detailed list of all the technical options available in the literature. To my knowledge, none of them ensure that the lines connecting projected areas do not cross, and so they do not guarantee the preservation of the lattice structure. However, they do improve visualization, and some of them address the issue of partitioning into "homonym" sub-trees.

7.4.2 *Toward a Geometric Embedding That Preserves the Lattice Structure*

Embedding of an Arborescence in a Lattice of Open Sets in a Topological Space. An approach that would guarantee the non-intersection of the projected lattice edges would consist of embedding arborescence A in a sub-lattice V of the lattice of open sets located in a topological space whose dimension is between 2 and 4 (for visualization) and that satisfies the following constraint: $\forall a, b \in A, \sup(a, b) = a \rightarrow p(b) \subset p(a)$, where p is the embedding of A in V. This would avoid the pitfalls of dimension reduction. Without providing a solution here, I might mention that I have found analogous objectives in the topological theory of graphs. Indeed, this theory considers embedding a graph's vertices in a topological space by associating a point to each vertex. If this embedding is possible, then in order to return to an embedding of open sets, it would suffice to associate each vertex with the envelope of the embedded vertices it subsumes. However, although algorithms for testing the planarity of a graph exist [7], i.e., if an embedding in a planar manifold exists, I am not aware of any algorithms that, beyond checking for the existence of such an embedding, actually enable its construction.

Scales, Fractals, and Wavelets. In order to achieve geometric embeddings that are isomorphic to multi-scale arborescences, one could also consider fractal mathematical objects. In particular, Sierpinski's fractals branch internally within a domain, a feature that echoes that of content-form arborescences, which themselves branch into internal details.[3] Yet another approach would be to draw inspiration from multi-scale wavelet decomposition.

In short, the geometric embedding of arborescences formalized as hypergraphs and lattices requires further investigation. As recalled previously, the arborescent paradigm takes precedence over the choice of an embedding. And although it does not offer the same visualization clarity, it allows us to account for lexical-semantic phenomena (Chap. 8) and to address composition (Chap. 9).

References

1. Fellbaum, C. (ed.): Wordnet. An Electronic Lexical Database. MIT Press, Cambridge, Massachusetts (1998)
2. Sagot, B., Fišer, D.: Building a Free French Wordnet from Multilingual Resources. In: OntoLex, Marrakech, Morocco (2008)
3. Fitch, W.M., Margoliash, E.: Construction of phylogenetic trees: a method based on mutation distances as estimated from cytochrome c sequences is of general applicability. Science **155**(3760), 279–284 (1967)
4. de la Barca, P.C., Camus, A.: La dévotion de la croix. Gallimard (2013)

[3] To obey the isomorphism between arborescence and fractal, one would have to be able, across the fractal set, to re-glue any lattice hyperedges that are disjoint but share descendants. Whenever re-gluing is possible, it could require embedding in the topological space of a higher dimension.

5. Van der Maaten, V., Hinton, G.: Visualizing data using t-SNE. J. Mach. Learn. Res. **9**(11) (2008)
6. McInnes, L., Healy, J., Melville, J.: UMAP: uniform manifold approximation and projection for dimension reduction (2018). arXiv:1802.03426
7. Hopcroft, J., Tarjan, R.: Efficient planarity testing. J. ACM (JACM) **21**(4), 549–568 (1974)

Chapter 8
Semantic Matter, Dynamics, and Basic Operators

8.1 Quasi-Fractal Arborescences as Units of Semantic Matter

I have defended the idea that the content form of a lexie is necessarily infra-linguistic, and that this follows from the fact that it connects two heterogeneous milieus, the expression milieu and the content milieu. I have also proposed that the lack of isomorphism between these milieus leads to a multi-scale construction obtained via the repeated effects of contrasts and similarities between signs. In previous chapters, I argued in favor of recourse to quasi-fractal, dual arborescences. In order to model them, I relied on a formalization in terms of hypergraphs, and in order to construct them, I made use of paradigmatic and opposition relations.

Arborescences are an alternative unit to symbols or vectors. They are designed to satisfy the requirements of adequacy for language phenomena such as polysemy, and to take into account the constitutive role of opposites. I use the term *semantic matter* to refer to these ever-changing units, which are renewed via the diversity of usages and diachronic changes, or throughout the learning process.

8.2 Operators and Dynamics

From a linguistic standpoint, syntax—whose syntactic part, for some authors, is inseparable from its semantic part—is the study of the natures and functions at play in a sentence. In written form, functions can be silent, or they can be marked, in which case they are attached to a lexeme, an affix, etc. Some functions act locally, as in cases of a noun qualifying an adjective, while others act at a distance. This is true of verbs, which distribute semantic-syntactic roles; it is also true of the functions associated with the logical conjunctions that structure discourse. In computational linguistics, the desire to provide a formal account of these various functions—among which I include the assignment of a grammatical nature—led to efforts to translate

S. Ploux, *Semantic Matter and Its States*, Lecture Notes in Morphogenesis,
https://doi.org/10.1007/978-3-032-22809-3_8

them into operators acting on formal units, whether symbolic, vectorial, geometric, or otherwise.

In order to distinguish between the language level and the modeling level in what follows, I will speak of *linguistic functions* and *formal operators*. Some authors, such as Culioli [1], included the notion of operator in their linguistic theory. However, since distinguishing between these levels is both fundamental and necessary, I chose here to pay particular attention to this distinction.

Vector models, then, are frequently equipped with addition, subtraction, convolution, and other operators that act upon vector units. These operators are used to formalize not only the computation of sentence meaning, but also similarities between units (words or sentences) and the search for shared semantic features [2]. In Pustejovsky's generative approach [3], operations such as coercion, co-composition, and selective binding determine the meaning of lexies in context. Each of these operations acts upon the attributes of a lexie or phrase, with the attributes themselves being related to the argumental, qualia-based, or event-related aspects whose value they specify.

Here, I propose operators that are suited to dual arborescences. I also incorporate dynamic aspects into my framework, not only the appearance and disappearance of lexies but also shifts in their semantic and syntactic boundaries. Although grammatical-structure changes over time are not addressed directly here, they are related to this. As in the physical sciences or even in the morphogenesis of living things [4], I hypothesize that the trajectory of a content form—its constitution, its contribution to sentence meaning, and its diachronic evolution over short or long periods - can be approached through the actions of operators (forces or fields in the natural sciences) exerted on semantic matter.

This chapter presents the simple operators that act upon the content forms of lexies and that can serve as a basis for modeling the functions of semantic conjunction. The operators used in constructing the parts of speech (PoS), as well as their conversions (from one PoS to another) and their functional roles in sentences, are described in the next chapter. Operators related to tense, aspect, or modality will not be addressed. In any case, the set of operators described here will suffice to establish an initial framework for semantic composition.

8.3 Some Examples of Dynamics: Creation and Evolution

8.3.1 Creation of Content Forms or "Word Making"

The pairing between an expression form and a portion of content is what creates a sign. For Saussure, as already mentioned, this involves a portion of content which in turn becomes inseparable from the expression form created at the same time.[1] In the structuralist approach, the uniqueness of the sign is governed by the interplay of

[1] In a certain way, the formation of a sign, and thus the process of putting the signified concept into words, reifies a piece of content in the language by associating it with a closed bounded unit in the space of expression forms. This holds true whether the associated content is interactive and

differences within the sign system. Complementary to this viewpoint, I consider here that while the emergence of the content form is indeed dependent upon differences within the sign system, it also possesses its own cohesion (its own internal structure) that that set of differences, when brought to bear on infralinguistic levels, allows us to describe. This internal structure is precisely what a dual paradigmatic arborescence attempts to model. Syntagmatic agreement, described in greater detail in the next chapter, completes the paradigmatic arborescence.

8.3.2 *Semantic Trajectory of a Lexie*

The meaning of a lexie can vary over time under various constraints such as those induced by competition among synonyms, a phenomenon widely described in the literature. Such diachronic processes can be initiated by external events whose effect is a redistribution of portions of content. In her thesis, A. Boussidan [5] described the short-term diachronic evolution of the French words *mondialisation* and *globalisation* (both can be translated into English as *globalization*). She formalized this through the computation of regular co-occurrence cliques (contextonymy) based on a corpus of journalism texts classified according to their chronological order of publication. The evolution of the resulting maps—each corresponding to a time window and constructed via the embedding of the contextonymy cliques containing the words being studied—provided evidence that societal events affected the evolution of the words' semantic attribution. This study showed how the word *mondialisation* shifted from a use referring to the opening of markets on a global scale to one referring to their political impacts. In contrast, its synonym, *globalisation*, remained confined to the economic sphere.

One could attempt to adapt this approach to the arborescence paradigm, since cliques are the arborescences's terminal elements. In such an adaptation, it would be necessary to jointly take into account paradigmatic relations and opposite relations in addition to syntagmatic agreement, whereas only regular co-occurrence was addressed in the above study. In this book, I have chosen to defer that topic and not detail the method for computing the evolution of content-form arborescences. To illustrate this computation without formalizing it, note that it would consist of either pruning or adding branches to the tree-like structure by making use of updates to substitution or permutation operations brought about by ever-changing usage. Tracing the dynamics of these shifts in meaning would help account for the lineage of meanings as portions or parts of a content form, whether aggregated, lost, or subtly modified.

---

dynamic, a spatial form, or a property with no parts, no contour, and/or no location. The grammatical noun is therefore the nature that is the most consistent with the word category.

8.4 Some Basic Operators Acting upon Arborescences

This section defines the basic operators acting upon lexical arborescences. Each definition is accompanied by some illustrative examples. Let $A_l = (A_{l^p}, A_{l^n})$ be the dual paradigmatic arborescence of lexeme l. A_l is a pair composed of arborescence A_{l^p} built from paradigmatic relations of semantic similarity, and of A_{l^n} built from antonyms, as described in Sect. 7.2.

Opposition Operator $\neg$

Operator $\neg$ is a generalization of the operator that associates lexie l with arborescence A_{l^p} the opposite arborescence A_{l^n}. Operator $\neg$ is applied to any arborescence A, whether or not it is derived from a single lexie, and returns $\neg A$ containing the cliques that are "antonyms" of arborescence A and are selected by matching (Fig. 7.2).

Merge Operator $\cup$

Let $\cup$ be the operator that associates the arborescence denoted $A \cup B$, generated from the set L of all lexies contained in A or B, with two arborescences A and B. The algorithm begins by computing the maximal cliques of L and the vertices of the hypergraph, and then, as described in the previous chapter, computes the smallest lattice containing them. Lastly, the dual arborescence of $A \cup B$ is computed (see Chap. 7). Figure 8.1 depicts the arborescence of $A_{aberrant^p} \cup A_{inutile^p}$ that combines the paradigmatic arborescences of the French words *aberrant* (aberrant or rogue) and *inutile* (useless). The opposite arborescence $\neg(A_{aberrant^p} \cup A_{inutile^p})$, too profuse to be shown here, includes the notions of "skill" (with markers like *ingénieux* (ingenious), *intelligent* (intelligent), etc.), "productivity" (with markers like *productif* (productive), *fécond* (fertile), etc.), and others based on correctness, reasonableness, and of course, usefulness.

Co-specification Operator $\cap$

Operator $\cap$ selects, in $A \cup B$, the hyperedges whose subsumed vertices do not belong exclusively to A or exclusively to B. This operator makes it possible to construct the arborescence—if it exists—that links A and B. Figure 8.2 depicts $A_{aberrant^p} \cap A_{inutile^p}$. The opposite arborescence (not shown here) calls more specifically upon the notions of *nécessité* (necessity), *efficacité* (efficacy or efficiency), and *importance* (importance).

Difference Operator $\triangle$

Operator $\triangle$ selects, in $A \cup B$, the hyperedges whose subsumed vertices belong to one and only one of the two composite arborescences. This operator makes it possible to generate the arborescence, if it exists, that differentiates between A and B and between B and A. It is complementary to operator $\cap$. Figure 8.3 depicts arbores-

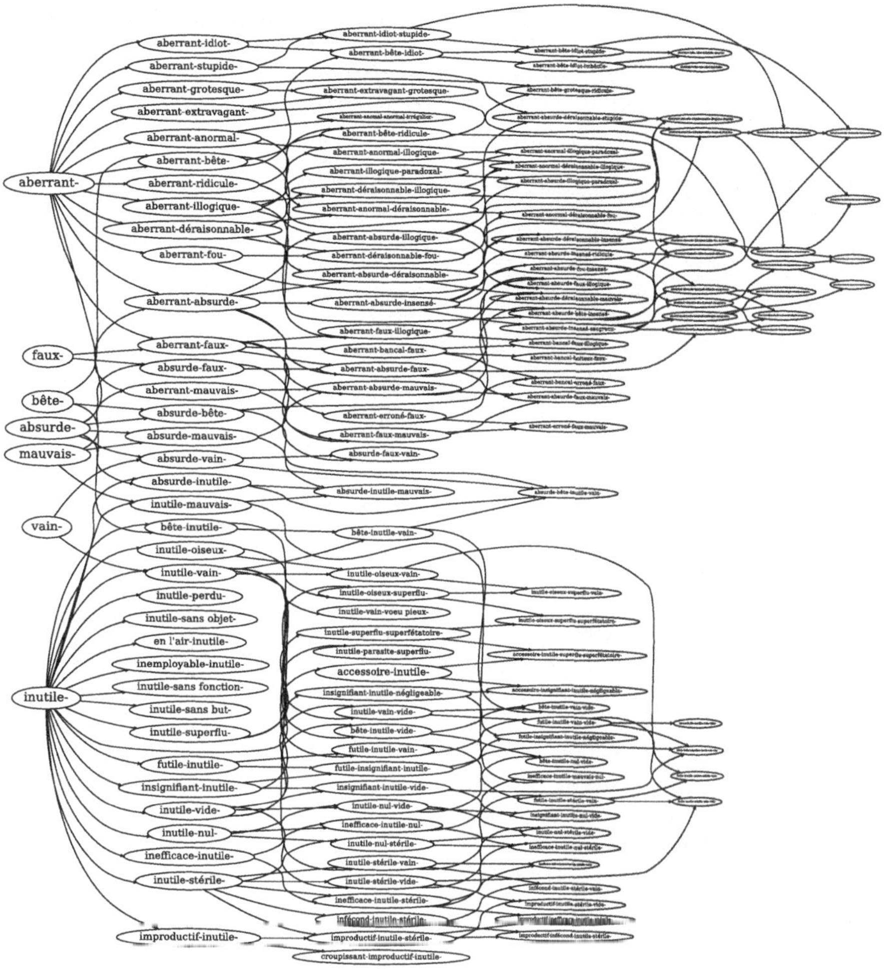

Fig. 8.1 Arborescence $A_{aberrant^p} \cup A_{inutile^p}$

cence $A_{aberrant^p} \triangle A_{inutile^p}$. The opposite arborescence (not shown here) contains, as its main markers, notions related to *productivité* (productivity), *logique* (logic), and *validité* (validity).

8.5 Basic Operators, Conjunction, and Semantic Isotopy

The basic operators (see above) provide a representation of a lexie's polysemy and its organization. They enable the computation of semantic sharing and connection between lexies, either directly (operator $\cap$) or by way of their opposite (operator $\neg$),

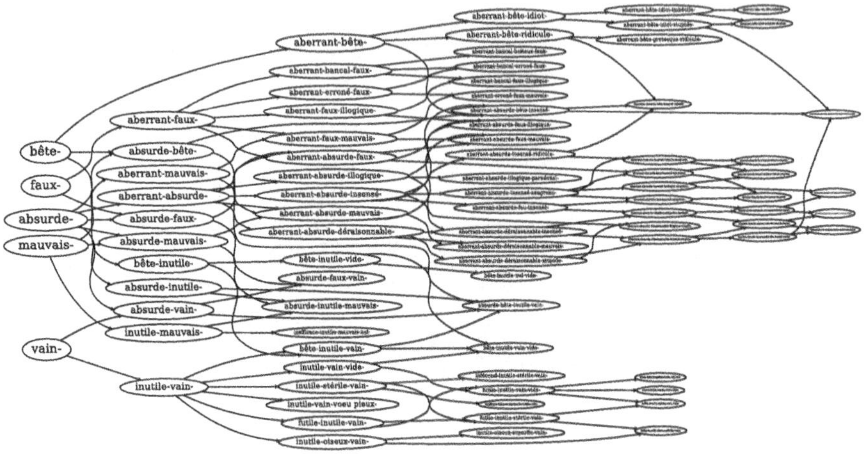

Fig. 8.2 Arborescence $A_{aberrant^p} \cap A_{inutile^p}$. The word *absurde* (absurd) is the main pivot point of the connection between *aberrant* (aberrant) and *inutile* (useless), with the words *faux* (false) and *vain* (vain), respectively associated to each one, as intermediaries

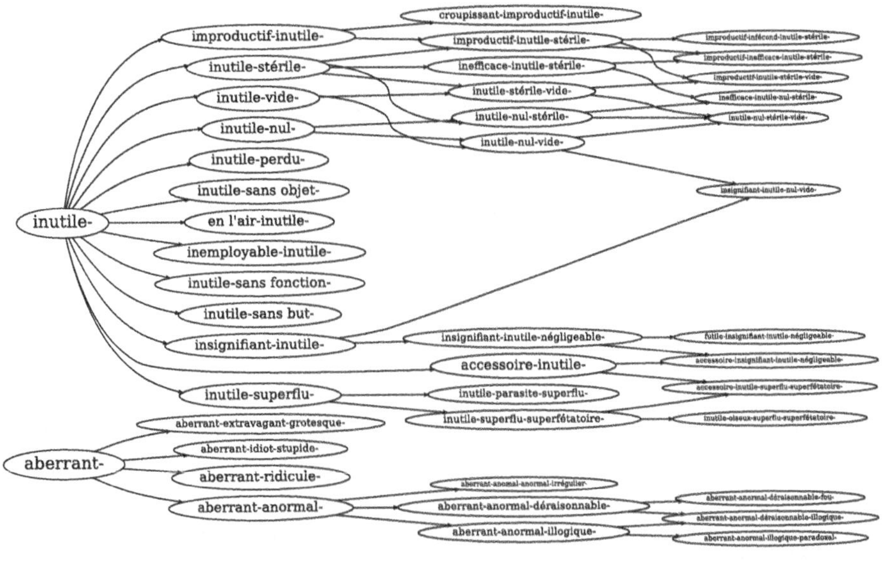

Fig. 8.3 Difference arborescence, $A_{aberrant^p} \triangle A_{inutile^p}$, between the paradigmatic arborescences of *aberrant* (aberrant) and *inutile* (useless). The main marker of their mutual differences (in this case, of higher order Ord) is the word *stérile* (sterile) for *inutile* (useless); for example, one might describe a discussion as sterile, or useless, because it has no positive outcome. The main marker is the word *anormal* (abnormal) for *aberrant*

or by their differences (operator Δ). They thus constitute an initial basis for studying and detecting semantic cohesion and isotopy within sentences and texts. In illustration, let us revisit the quote from Chateaubriand (Chap. 2) in its context.

Cet étroit espace me parut propre à renfermer mes longues espérances [...] Les arbres que j'y ai plantés prospèrent, ils sont encore si petits que je leur donne de l'ombre quand je me place entre eux et le soleil [...] ils rappellent mes voyages et nourrissent au fond de mon cœur d'autres illusions.

This narrow space seemed suitable for containing my long-held hopes [...] The trees I have planted here are thriving; they are still so small that I shade them when I stand between them and the sun [...]. They remind me of my travels and, deep within my heart, they nourish other illusions. (our translation)

The word groups *étroit* (narrow), *long* (long), and *petit* (small), in red; *espérances* (hope) and *illusion* (illusion), in blue; and *ombre* (shade) and *soleil* (sun), in green are composed of lexies that share portions of their paradigmatic, opposite, or composite arborescences. This sharing of parts indicates semantic cohesion within the paragraph, as computed here.

All of the operations presented here are based on lexical relations within a given category. However, they could easily be generalized by taking category conversions into account (e.g., *young* $\leftrightarrows$ *youth*). To do so, it would suffice to use a two-category conversion relation to match the arborescences specific to each one, as presented earlier for the antonymy relation (Fig. 7.2).

8.6 Basic Operators and Semantic Interpolation

Beyond the study of the semantic isotopies that run throughout texts, applying operators to arborescences makes explicit a semantic implicit that fills in the space between the content forms. I call this process "*semantic interpolation*".

Take the example of the French expression *aller et venir* (to come and go, see note 2). It contains two opposing verbs with the coordinating conjunction and in-between, which implies a temporal sequence[2]. However, in addition to the sequencing of actions, the expression also implies notion sharing. Thus, the opposite arborescence of the one constructed from the co-specification operator $\neg(A_{aller^p} \cap A_{venir^p})$ has—as the highest order elements ORD—the verbs *rester* (stay) and *demeurer* (remain), as well as *fuir* (flee), *quitter* (leave), and *disparaître* (disappear), which state in the

[2] In French, when the conjunction *et* (and) connects adjectives, it most often refers to notions that reinforce each other, such as *doux* (sweet) and *gentil* (nice), or *riche* (rich) and *puissant* (powerful). When pertaining to nouns, *et* usually connects complementary words, as in *père et mère* (father and mother) or *ciel et terre* (heaven and earth). Finally, when applied to verbs, *et* most often expresses a temporal succession, as in *écrire et signer* (write and sign). In this last case, it is equivalent in usage to the adverb *puis* (then). Sometimes, *et* links verb opposites, as in the l'expression *aller et venir* (go and come, whose equivalent English expression is "come and go"), where two aspects (sequentiality and opposition) are at play.

negative what *aller et venir* does not mean, namely, the absence of movement or, on the contrary, a movement without a return[3]. This example stresses that the result of the semantic interpolation follows from the topology induced by the cliques' granularity and from matching via lexical relations (here, antonymy). It also points out that ultimately, the discovery of the main markers of this interpolation rests on the arborescent structure formalized using the notions of hypergraph and lattice.

References

1. Culioli, A.: Pour une linguistique de l'énonciation: opérations et représentations. Editions Ophrys (1990)
2. Mikolov, T., Sutskever, I., Chen, K., Corrado, G.S., Dean, J.: Distributed representations of words and phrases and their compositionality. In: Advances in Neural Information Processing Systems, pp. 3111–3119 (2013)
3. Pustejovsky, J.: The Generative Lexicon. MIT Press, Cambridge, Massachusetts (1998)
4. Thompson D'Arcy, W.: On Growth and Form. Cambridge University Press **1**(6), 7 (1942)
5. Boussidan, A.: Dynamics of semantic change: detecting, analyzing and modeling semantic change in corpus in short diachrony. Ph.D. thesis, Universit Lyon II (2013)

[3] Here are two examples of cliques matched by the antonymy relation to the cliques of the arborescence of $\neg(A_{aller^p} \cap A_{venir^p})$ ($\neg(go \cap come)$): (1) *demeurer, rester, s'arrêter, s'attarder, stationner, séjourner* (remain, stay, stop over, linger, park, stay over); (2) *abandonner, disparaître, fuir, partir, quitter, s'enfuir, s'éloigner* (abandon, disappear, flee, leave, depart from, run away, go away).

Chapter 9
Syntagmatic Arborescences and Composition Operators

Semantic composition can be defined as the transformation of content forms within a sentence (or phrase), transformations that are partially achieved by assigning the syntactic natures and roles induced by the sentence. Reciprocally, but to a lesser extent, the content forms constrain the assignments. From this process, a global coordinated meaning emerges. To account for this, it is necessary to determine what these syntactic natures and roles encompass from the linguistic, neurolinguistic, and psycholinguistic standpoints (see Chap. 3), in order to then state the possibilities and limitations of a model grounded in the expression form. To this end, I will begin here by sketching out what aspects of content may turn out to be inaccessible in a one-directional approach. Then I will present the operators responsible for the effects of syntactic roles on semantic composition and how those operators are rendered in the arborescence paradigm.

9.1 The Semantics of Syntactic Natures and Roles

Based on neurophysiological studies, as well data on usage in the language and theories of child development, I have attempted to highlight cases where syntactic natures and roles exhibit a strong acquaintanceship with cognitive content and grounding (Chap. 3). I have also reminded the reader of the existence of a massive phenomenon of part-of-speech conversion—such as noun-to-verb conversion and vice versa, or noun-to-adjective conversion and vice versa. Two lessons can be drawn from these elements. The first is that natures and roles, like lexies, have a content form that links them to nonlinguistic content. The second is that there are grammatical assignment and conversion operators that act on semantic matter in specific and differentiated ways. However, the expression-form/content-form duality, when assigned to grammatical natures and roles, lacks the richness found in so-called "full" words.

S. Ploux, *Semantic Matter and Its States*, Lecture Notes in Morphogenesis,
https://doi.org/10.1007/978-3-032-22809-3_9

Parsimony of the Sign System Assigned to Function Words and Markers. While the number of full words in any language is considerable, this magnitude contrasts with the parsimony of grammatical markers (morphology, syntax), whether expressed or implied by word order. The number of markers characterizing the parts of speech (verb, noun, adjective, etc.) is small (eight for French) or nonexistent in certain languages. What's more—again using the example of French—verb lemmas include few endings (e.g. *-er*, *-ir*, *-re*), which are sometimes shared with those of other parts of speech. For instance, a French noun can end in "er" as in *plancher* (floor), "ir" as in *plaisir* (pleasure), or "re" as in *cadre* (frame). In addition, the number of inflected forms in French, although higher than in other languages like English and *a fortiori* Chinese, remains limited. For French again, the scarcity of distinctive markers of nouns or adjectives accentuates this overall phenomenon. Other languages—such as case languages where syntactic functions can be marked morphologically—have a richer inventory of syntactic expression forms, albeit succint.

To conclude this brief overview, noun determiners, not only verb aspect markers (such as the Mandarin character 了 (le in Pinyin), but also the perfective particle that often indicates the completion of the verb), and all such function words that are dislocated from the lexie to which they pertain, are highly frequent in the usage of languages that have them, but they, too, are few in number.

Implications of This Parsimony on the Semantic Span of Content Forms. It follows from this general parsimony that both the content form of function words and the grammatical units of the language are anchored in a limited number of expression forms and are linked to contents as rich as those derived from temporality, spatiality, reification, interaction, qualia, and the process of converting them from one to the other, etc. A corollary of this disparity is a loss, in terms of expression, of what the associated content covers. Thus, the valence of verbs—which may or may not be marked in a language-specific way by particles, endings, or word order—is very reductive in the light of the different types of processes and interactions expressed.

As a result, the use of a method based on repeatedly applying a set of differences and similarities between expression forms, as presented earlier, seems ill-suited to constructing purely grammatical content forms. So my proposal is to rely on the combination of assigning grammatical natures and roles, and assigning lexical content forms. In short, the idea is to make use of syntagmatic agreements, which are no longer understood here to mean co-occurrence profiles but structures shaped by syntactic patterns in acquaintanceship with paradigmatic relations.

This chapter begins with a presentation of how the syntagmatic arborescences associated with the main syntactic natures and roles are constructed. First, each nature and each role will be introduced in order to find out what part of its associated content is left out when the model used is a model that goes in the expression-form to content-form direction. To do this, I will revisit some elements from the neurolinguistic, linguistic, and modeling work presented above. The part on modeling will be situated with respect to Thom's dynamic model, which, in my mind, is the best suited to highlighting both how semantic content cannot be attained using the arborescence paradigm and how that paradigm is better suited to representing content forms.

Once the limitations are stated—just as with dual paradigmatic arborescences—I will present a construction operator. The construction process is based here on the pre-compilation of usage examples detected in a corpus of literary texts.[1] This process could be reproduced using any other corpus, such as corpora containing different text genres, whether journalistic, scientific, essays, etc.

The corpus having been chosen, occurrences of the most frequent syntactic patterns based on the respective frequencies of the lexemes[2] are extracted. Figure 9.1 gives a schematic illustration of how regular usage detected in the corpus leads to the construction of a syntagmatic arborescence. Next, I will show how this approach is adapted to the different parts of speech and to the subject and object functions for the noun category. This process is regular and can be extended to all syntactic patterns. The illustrative sentence, *Cet étroit espace renferme mes longues espérences* (This narrow space contains my long-held hopes (our translation)), taken from Chateaubriand's Mémoires d'outre-tombe and simplified for the purpose of this presentation, will serve as a example and will be revisited later in this chapter, where compositionality effects are addressed.

9.2 Syntagmatic Arborescences

9.2.1 Verbs

As neurophysiological data has shown and as proposed by Thom, the content of the verb category is fundamentally dynamic. In catastrophe theory, verb dynamics take place in interaction scenarios between protagonists, which themselves are likened to points. Thom describes the semantics of a given verb as a curve within the space of control parameters resulting from an unfolding operation. This continuous, spatialized conception, albeit organized around catastrophes corresponding to singular moments in verbal dynamics, differs in essence from the temporality induced by language, because, as several authors have shown (see for example [1]), the linguistic temporality of verbs is not spatial, it is directional and constitutes an irreversible change, whereas spatiality is by nature reversible.

Moreover, linguistic temporality does not unfold within a continuous and homogeneous dynamic that would be as uniform as the countdown of a clock, or the equally uniform format of a one-dimensional manifold that would go from an earlier configuration to a later configuration due to its having passed through a particular point in time. On the contrary, unlike an amount of time that is monotonic and exter-

[1] The corpus used contains 321 French novels from the 19th and 20th centuries, made accessible by Project Gutenberg https://gutenberg.org.

[2] Here, Pointwise Mutual Information (PMI) is used as the frequency index of co-occurrence within a given pattern.

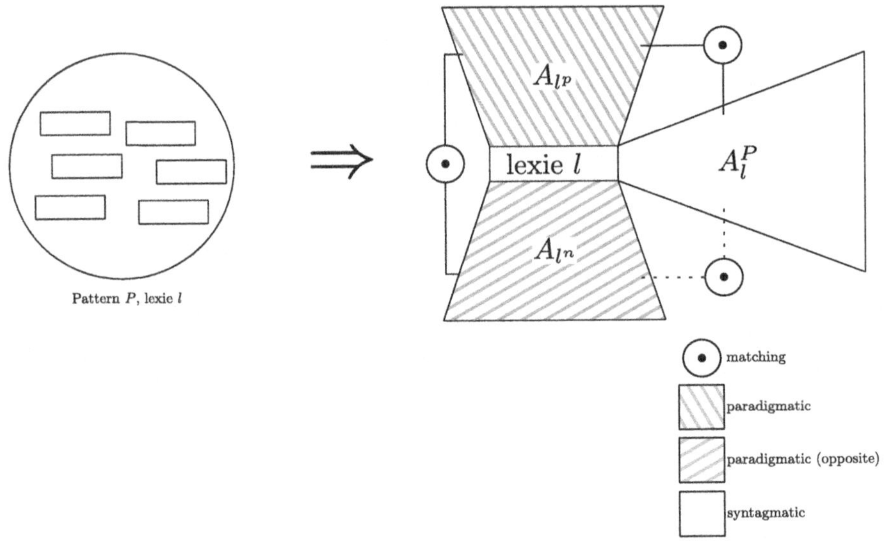

Fig. 9.1 On the left: results of an extraction from a corpus of regular occurrences of syntactic pattern P around lexie l. For example, if the pattern is $adj + l$, where adj is an adjective modifying l, the system extracts from the corpus those adjectives that occur regularly in this pattern. Regularity is defined here based on the frequencies of adj and l, their co-occurrence frequencies, and their co-occurrence distributions in the corpus. On the right: construction of the dual paradigmatic arborescence $A_l = (A_{l^p}, A_{l^n})$ of lexie l and of the syntagmatic arborescence A_l^P associated with pattern P. The syntagmatic arborescence results from several steps. The first three are (a) computation of the arborescence $A_{l^p}^P$ containing the paradigmatic cliques of the elements of pattern P (in the example of pattern $adj + l$, these are the extracted adjectives modifying one of the terms in the paradigmatic arborescence A_{l^p}); (b) computation of the arborescence $A_{l^n}^P$ containing the paradigmatic cliques of the elements of pattern P when it includes a term from the opposite arborescence $\neg(A_{l^p}) = A_{l^n}$; (c) computation of the arborescence $A_l^P = A_{l^p}^P \cup (A_{l^p}^P \cap A_{l^n}^P)$. This procedure takes into account the opposites of lexie l, if they exist, provided that the regular occurrences of pattern P that include them, and the resulting arborescence $A_{l^n}^P$, are connected via operator $\cap$ to those of A_l^P. In a second phase, the paradigmatic cliques of A_l are matched to the syntagmatic cliques of A_l^P using the method defined above in Sect. 7.2, via the relation "forming a regular occurrence of syntagmatic pattern P that combines terms from A_l with terms from A_l^P". Only those syntagmatic cliques matched to at least one clique of the paradigmatic arborescence A_l are selected. The syntagmatic arborescence, as defined in Chapter 7, is the smallest lattice for the intersection and the union containing the selected cliques

nal to the processing of the verb, specific positions or time windows are marked in language. There are three cases that underline the difference between temporality expressed in language and mathematical, spatialized time: (1) the intrinsic duration of the verb, which opens a time window on the action (*yesterday, he was chopping wood*); (2) the gathering up of the process into a single instant (*he chopped wood then disappeared*); and (3) the situating of the process, either in relation to when the utterance was produced (*tonight, I'll chop wood*) or to a point in time marked in the utterance (*when I arrive, he will have chopped the wood*).

This divergence (in terms of temporality) between language and a continuous spatialized model must not mask what links them. The verb seen as a curve in a morphodynamic landscape that organizes the type of interaction, or the verb seen as the organizing center (nucleus) of the sentence and medium for assigning the subject and complement roles, are partly homologous aspects. The governance of the interaction dynamic at the content level is projected onto the lexie that carries the sentence's verb and is materialized in the governance of its structure. It is precisely the role-distribution function that the operator associated with the verb will fulfill. Through this governance process, the verb will appear as a role category, i.e., a category defined by reference to a process, a role defined by the governance it draws from the morphodynamic interactive landscape to which it refers and from the distribution of the subordinate roles that governance involves.

9.2.2 Syntagmatic Arborescences of Verbs

Let υ be a lexie, A_υ its dual arborescence, and V the operator which, when applied to υ, generates the syntagmatic arborescences associated with the grammatical subject of υ (denoted $A_\upsilon^{s-\upsilon}$) and with its complements (denoted $A_\upsilon^{\upsilon-c}$) (provided the process in question has a positive valence).

The syntagmatic arborescences associated with each of these argument roles—subject, direct object, and so on—are constructed by following the procedure defined above in Fig. 9.1 for the Subject-Verb pattern and the Verb-DirectObject pattern, respectively. Once the subject and direct-object arborescences are constructed, they are matched by way of the paradigmatic arborescence A_υ (Fig. 9.2). This interaction between the syntagmatic arborescences of the verb's arguments, lacking in the tree models of generative grammar and in Tesnière's stemmas, brings the arborescence model presented here closer to Thom's elementary interaction graphs.

The rest of the figures (Figs. 9.3–9.7) present the results of the different steps for the French verb *renfermer* (to contain or enclose). In Step 1 (Figs. 9.3 and 9.4) the dual paradigmatic arborescence of *renfermer* is constructed. In Step 2 (Figs. 9.5 and 9.6), the syntagmatic arborescence of subjects for which *renfermer* is the verb is constructed, in two substeps. In Step 2' (Fig. 9.7), the syntagmatic arborescence of direct objects governed by *renfermer* is constructed. Lastly, in Step 3, the argument arborescences are matched by the pattern relation S-V-O. A few examples of the matching results are listed in Fig. 9.8.

9.2.3 Nouns *and Their Roles*

Thom likened the noun to a ball in a multidimensional space,[3] and by contraction, to a point. One could subsequently propose that nouns are anchored in the possibility

[3] "As a general rule, each of the actants (of the basic sentence) is topologically a (contractile) ball. This is the case, for example, of animate (living) things. At each instant, we shrink each of the actants into one of its points..." [7].

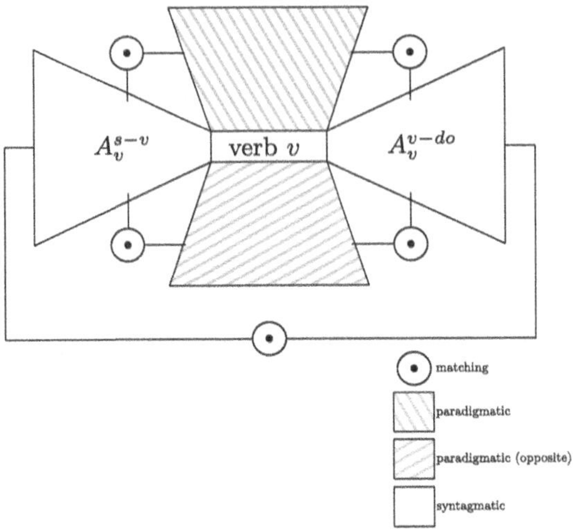

Fig. 9.2 Illustration of syntagmatic arborescences associated with verb v and its dual paradigmatic arborescence A_v. Each of the syntagmatic arborescences is constructed using the method defined above in Figure 9.1. First, the subject arborescence is based on regular uses of the Subject-Verb pattern extracted from the corpus (likewise for the syntagmatic direct-object arborescence). The arborescences of the verb's arguments are then matched. For this, a clique c^S of the syntagmatic Subject arborescence is matched to clique c^C of of the syntagmatic Direct-Object arborescence if there exists a clique c_v in A_v such that c_v is matched to the subject clique c^S by the Subject-Verb pattern and to clique c^C by the Verb-DirectObject pattern. Arborescences A_v^{s-v} and A_v^{v-do} result from the matching

of reification by delineation of a spatializable form. It is thus hypothesized here that substantivizing amounts to confining, inside an enclosure or envelope, all of the internal characteristics and dynamics of the substantivized source, and/or all of the external characteristics and dynamics of the substantivized target, based on all occurrences of uses and interactions.

Before continuing, I would like to stress that attributing to a noun a spatializable entity that is both the source and the object, each with its own specific dynamics, does not account for all nominalization phenomena. Indeed, like verbs, some nouns, called relational nouns or predicate nouns, contain the mark of a link, a relationship, or a process between entities. This is the case, for example, of ties of father kinship (being a father is being the father of someone). More generally, this is the case of all nouns that necessarily (although sometimes implicitly), invoke another entity (e.g., *a friend* or *an enemy*). Several ways of describing this phenomenon have been studied (see [2]). For the noun category, is it an encapsulation of a process that binds the protagonists? Under this hypothesis the word *friend*, for example, would contain (within it) a process (such as being fond of) that connects the source and target actants of that affection. I will not address this potential hybridization of categories, which would require additional development. Instead, I will concentrate on the characteristics that are the most specific to verbs and nouns.

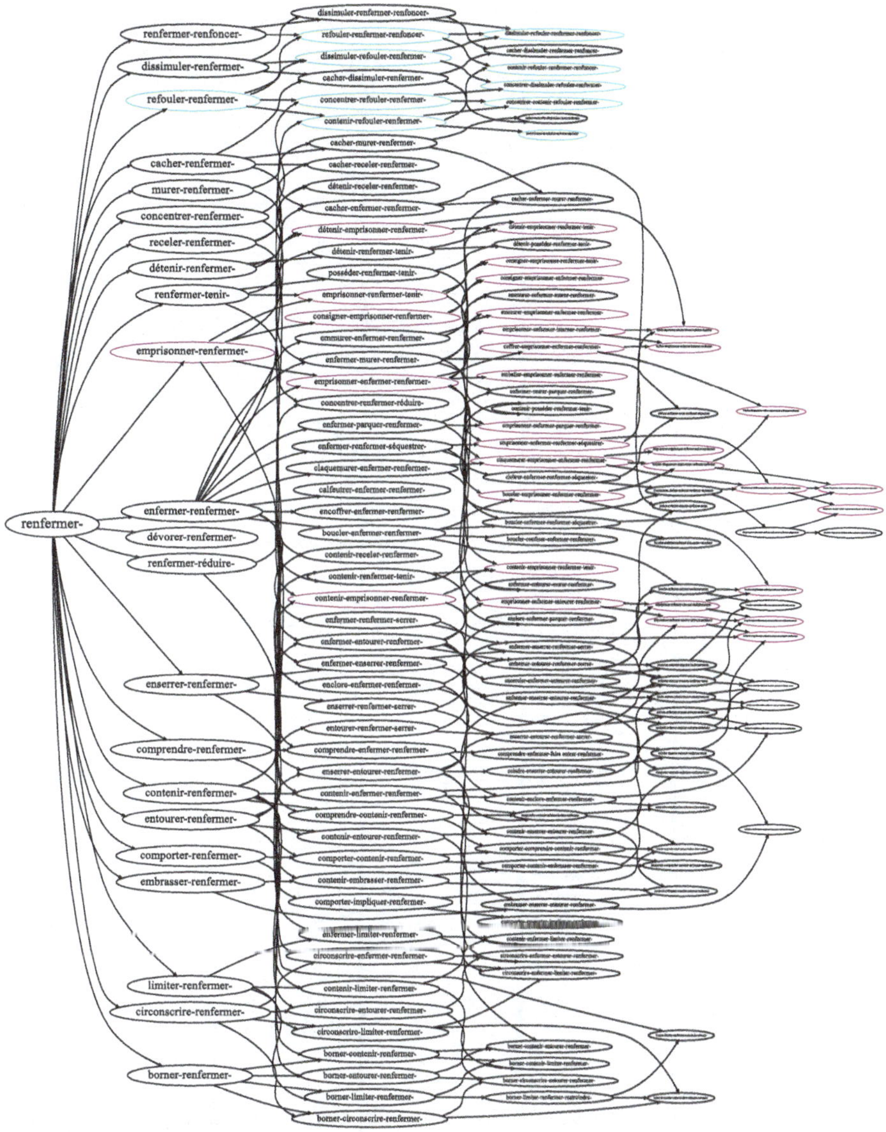

Fig. 9.3 Step 1 (a): Paradigmatic arborescence $A_{renfermer^p}$ of the verb *renfermer* (to contain, to lock, to confine). The hyperedges containing *emprisonner* (to imprison), for example, are encircled in magenta; those containing *refouler* (to repress, to contain), are encircled in cyan

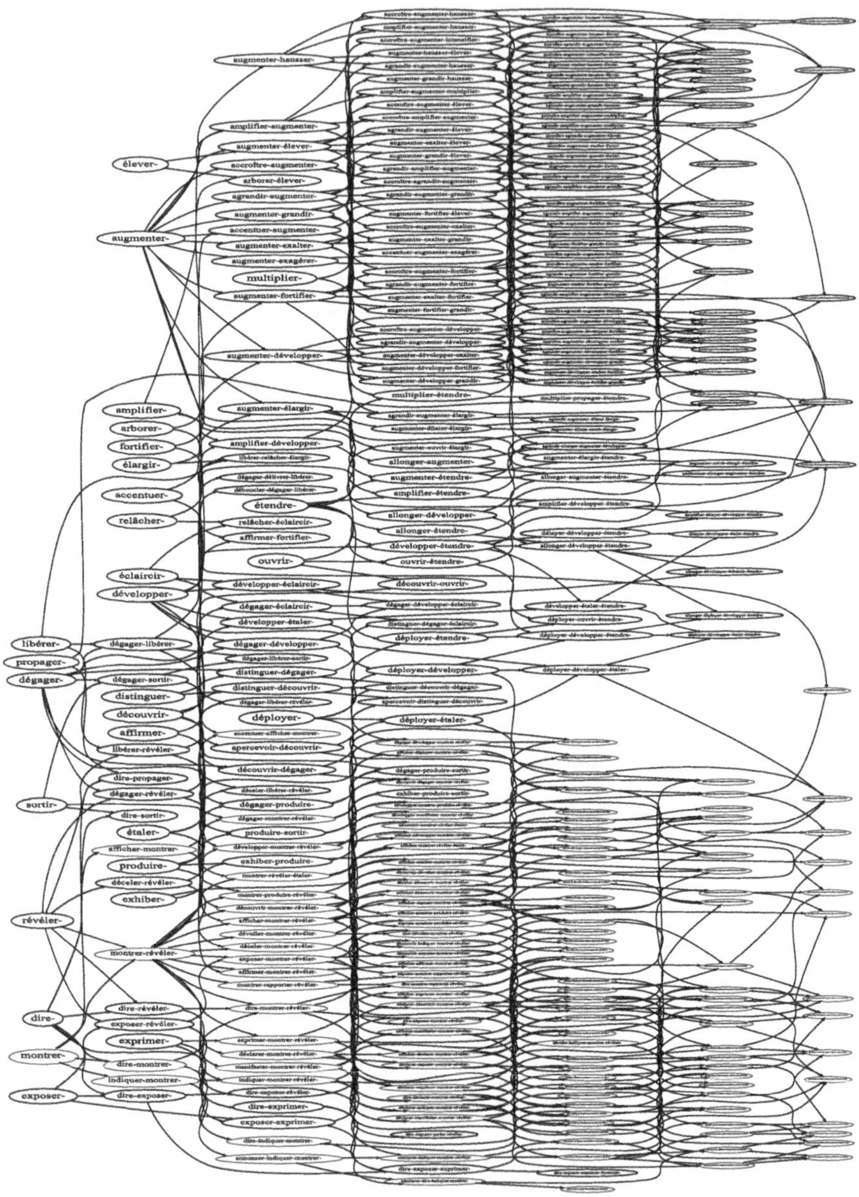

Fig. 9.4 Step 1 (b): Paradigmatic arborescence $A_{renfermer^n}$, the opposite of $A_{renfermer^p}$. The hyperedges containing *montrer* (to show), for example, are encircled in magenta

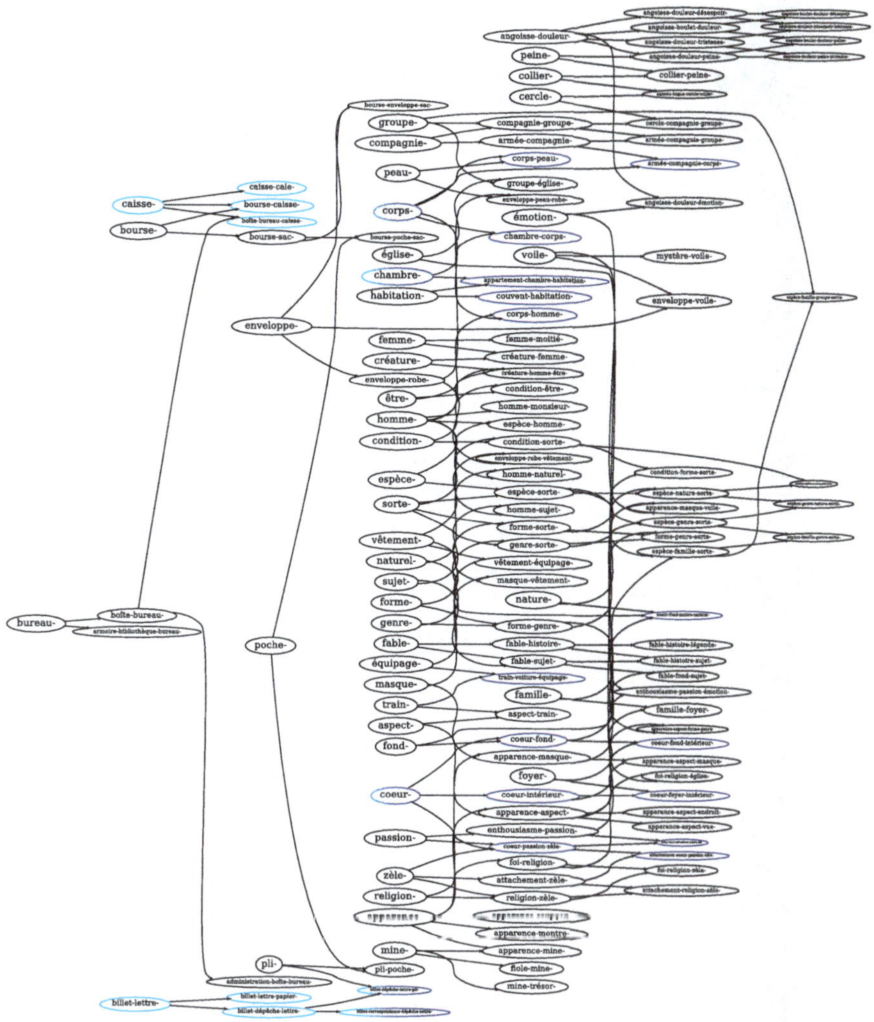

Fig. 9.5 Step 2 (a): Arborescence of the regular subject lexemes extracted from the corpus and constructed from arborescence $A_{renfermer^{p}}$. It includes lexies that are containers themselves, such as *enveloppe* (envelope), *caisse* (crate), and *armoire* (wardrobe), as well as lexemes that have the potential to be containers, like *letter* (letter, as in "the letter contains fragments of family life"), or *cœur* (heart, as in "his heart contains a tormented soul"; "the heart contains the vena cava"

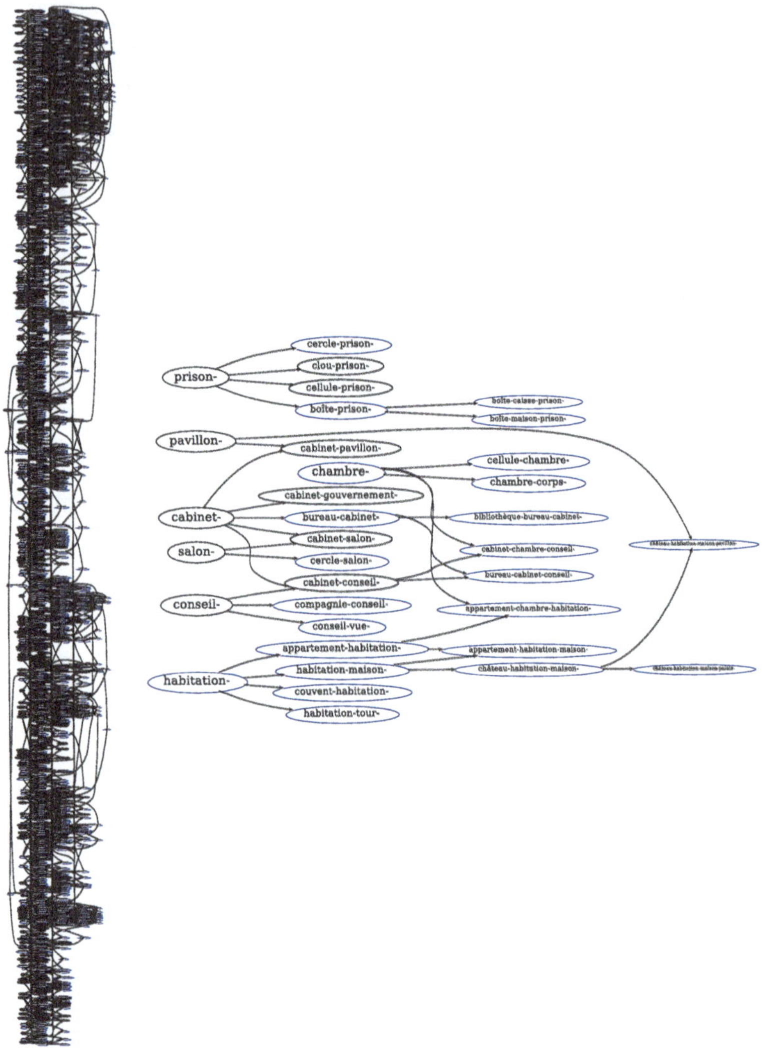

Fig. 9.6 Step 2 (b): On the left, we see the arborescence $A^{s-v}_{renfermer} = A^{s-v}_{renfermer^p} \cup (A^{s-v}_{renfermer^p} \cap A^{s-v}_{renfermer^n})$ of regular subject lexemes extracted from the corpus and constructed from arborescence $A_{renfermer^p}$ and its opposite $A_{renfermer^n}$. The blue hyperedges are those containing a lexeme from $A^{s-v}_{renfermer^p}$. Here, the use of opposites (by way of $A^{s-v}_{renfermer^n}$) densifies the results obtained and adds new candidate subjects. Indeed, although $A^{s-v}_{renfermer}$ is more complex and denser than $A^{s-v}_{renfermer^p}$ presented above, they share many lexemes that are distributed uniformly. On the right, we see a zoomed-in portion of the arborescence containing nouns referring to assemblies (*cabinet* (cabinet), *chambre* (chamber), *conseil* (council)), or to places for (*habitation* (dwelling), *pavillon* (building)). Note here that words like *prison* (prison), *salon* (living room), and *pavillon* (building), in black), show up as potential subjects of the word *renfermer* (to contain), even though they do not appear as subjects of *renfermer* in the corpus nor in $A_{renfermer^p}$

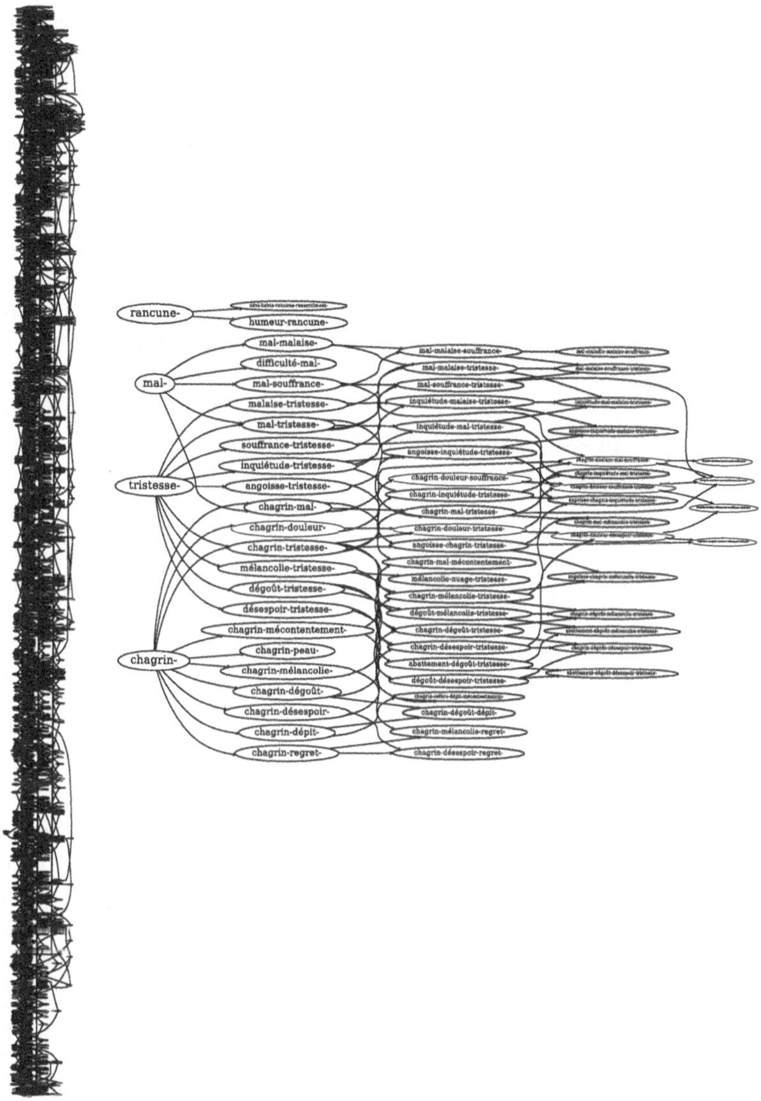

Fig. 9.7 Step 2' (b): On the left, we see the arborescence $A^{v-do}_{renfermer} = A^{v-do}_{renfermer^p} \cup (A^{v-do}_{renfermer^p} \cap A^{v-do}_{renfermer^n})$ of regular direct-object lexemes extracted from the corpus and constructed from arborescence $A_{renfermer^p}$ and its opposite $A_{renfermer^n}$. This arborescence is quite large due to the variety of lexemes serving as a possible direct object of the verb *renfermer*, or more generally, of any verb in its paradigmatic arborescence $A_{renfermer}$. Indeed, in a certain sense, almost anything can be *renfermé* (contained or enclosed) within a container. Nonetheless, certain usages are more characteristic of the literary genre, such as *richesse* (wealth) or *trésor* (treasure). While in subject position $(A^{s-v}_{renfermer})$, we find abstract lexemes like *apparence* (appearance) or *aspect* (aspect), evoking externality, in direct-object position, we find lexemes related to particular emotions like *chagrin* (grief) and *rancune* (resentment), which evoke internality. On the right, a zoomed-in view of a portion of the arborescence shows some examples

Cliques of Arborescence $A^{s-v}_{renfermer}$ (subject) of the verb *renfermer*		Cliques of Arborescence $A^{v-do}_{renfermer}$ (direct object) of the verb *renfermer*
armoire ; bibliothèque ; bureau (wardrobe, bookshelf, desk)	↔	album, cahier, livre, registre (album, notebook, book, register)
billet, carte, lettre (note, card, letter)	↔	demande, désir, vœu (request, desire, wish)
carnet, journal, mémoire (notebook, diary, memory)	↔	existence, histoire, vie (existence, story, life)
air, attitude, expression, figure, physionomie, trait, visage (air, attitude, expression, face, physiognomy, feature, face)	↔	courage, force, volonté, âme, énergie (bravery, strength, willingness, soul, energy)

Fig. 9.8 Step 3: Match between the syntagmatic arborescences $A^{s-v}_{renfermer}$ and $A^{v-do}_{renfermer}$. For each of these two arborescences, this figure shows a few examples of pairs of cliques matched on the basis of the relation induced by the Subject-Verb and Verb- DirectObject patterns as they move through arborescence $A_{renfermer}$

9.2.3.1 Facets and Roles of Nouns

Thus, the grammatical functions carried by a noun in a sentence expressed in the active voice rely on the distinction between the source dynamic and the target dynamic. In this approach, the grammatical subject is the benchmark, i.e., the source of the interaction referred to by the verb, and the verb is a potentiality of the subject. The direct object is the target of the pair formed by the verb and the subject from which it proceeds.[4] I thus propose that the Subject operator brings up an arborescence of all verbs that deploy the potentialities of the noun as the source of a verb, and that the DirectObject operator is the operator that assigns, to a noun, an arborescence of the verbs that govern it. The focus here is on the category and role of the noun within the kernel sentence Subject-Verb-DirectObject. Other syntagmatic operators of nouns, such as those associated with a noun complement or its modifiers can, in the same manner, be expressed by syntagmatic arborescences. The example of noun modifying is developed below.

[4] Remember that language can be freed from this original semantic pattern (see Chap. 2), which remains syntactically structuring. So in French as in English, for example, the subject or object position can be indifferent for certain verbs, as in *Les prix augmentent* (the prices increase) vs *augmenter les prix* (increase the prices) refers here, in both cases, to prices that are increased and whose original instigator (i.e., the subject, a shopkeeper for example) may or not be specified.

9.2.4 *The Syntagmatic Arborescence of* Subjects

Let n be a lexeme, A_n its dual arborescence, and S the operator that generates the syntagmatic arborescence A_n^{s-v} assigned to the Subject-Verb pattern, with n standing for the subject. The algorithm follows the process described in Fig. 9.1. A_n^{s-v} contains the most prevalent active verbs in the corpus whose subject belongs to the paradigmatic arborescence A_{n^p}.[5] Continuing with the illustrative sentence chosen above, the noun subject *espace* (space) has a very small syntagmatic arborescence A_{espace}^{s-v}. Indeed, in the corpus used here, *espace* is never in the subject position. As in the case of the syntagmatic arborescences of the verb *renfermer*, and in general when the terms' dual paradigmatic arborescences are taken into account, the syntagmatic arborescence constructed contains potential candidates. In our example, then, we find the French verbs *couler* (flow), for which examples absent from the corpus were detected, as in "Les espaces coulent, presque sans limites, la transparence est une constante" (Spaces flow, almost endlessly, transparency is a constant; our translation) [3]), and *circuler* (circulate), as in "La litterature offre un autre partage du paysage où les espaces circulent et se superposent, ou l'intime et l'historique se répondent" (Literature offers another sharing of the landscape where spaces circulate and overlap, where the intimate and the historical respond to each other; our translation) [4]).

9.2.5 *The Syntagmatic Arborescence of* Complements

Let n be a lexeme, A_n its dual arborescence, and C the operator that applies the process described in Fig. 9.1 to generate the arborescence A_n^{v-c} associated with n used as a complement. Figure 9.9 shows the results for the noun *espérance* (hope), with the Verb-DirectObject structure as the pattern.

9.2.6 *Modifiers and Their Role*

Adjectives and adverbs modify the lexie or phrase to which they apply, without changing its category or role (e.g., "a magnificent painting", "leave quickly"). While nouns are often organized according to an inclusion order and verbs follow a temporal order, modifiers are governed by a gradient. The French usage "bible", Le Bon usage [5], speaks in terms of degree in such a way that even when the modifier comes from a noun (e.g., square), it is possible to graduate its value ("a table that is rather square"). The gradient allows for an abundance of antonyms. As with verbs, where antonymy most often consists of reversing the temporal dynamic (to go–to come), antonymy for adjectives consists of reversing the degree or intensity (tiny–small–big–immense). The nature of the order that applies to nouns also explains why nouns have fewer

[5] It would also be possible to add passive-voice verbs that have n as the object.

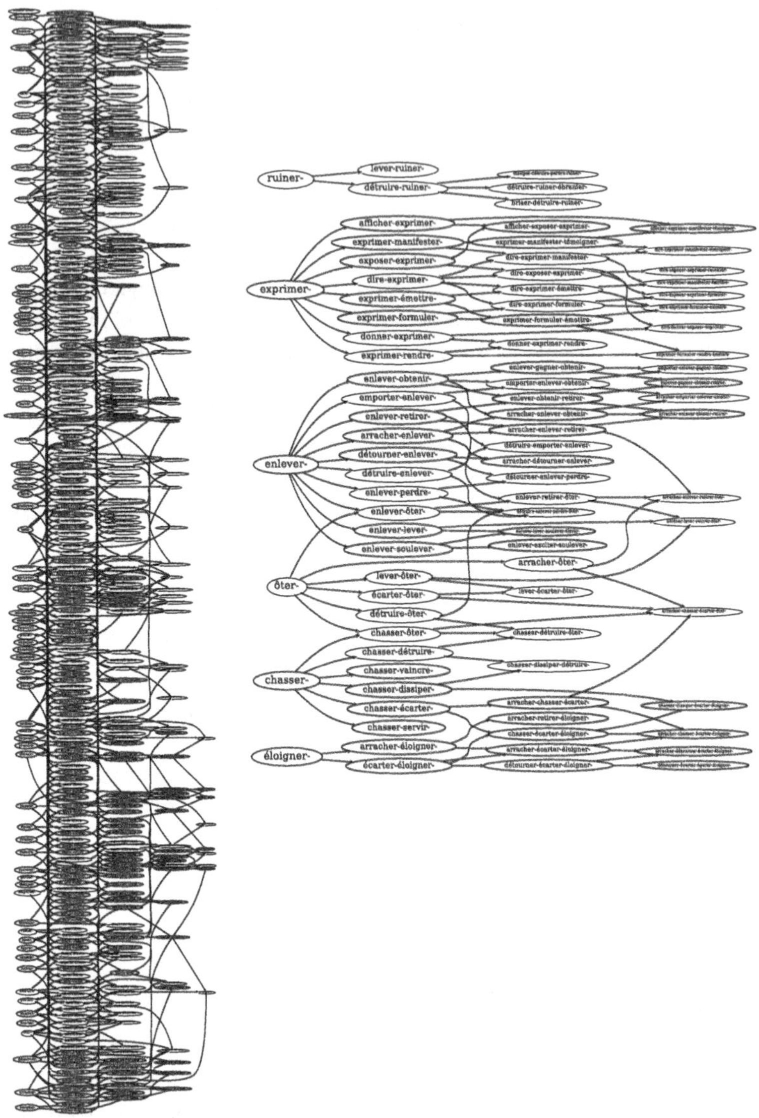

Fig. 9.9 On the left, we see an overview of arborescence $A^{v-do}_{espérance} = A^{v-do}_{espérance^p} \cup (A^{v-do}_{espérance^p} \cap A^{v-do}_{espérance^n})$ of regular verb lexemes extracted from the corpus and constructed from the paradigmatic arborescence of *espérance* (hope) $A_{espérance^p}$ and its opposite $A_{espérance^n}$. It mainly includes verbs related to loss, such as *perdre* (to lose), *ruiner* (to ruin), *ébranler* (to shake up); accomplishment, such as *accomplir* (to accomplish), and *combler* (to fulfill); and expressing, such as *dire* (to say), *exprimer* (to express), and *avouer* (to confess). The blue hyperedges are those that contain one of the verbs for which *espérance* (hope) regularly serves as a direct object in the corpus. The black hyperedges are inferred computationally. On the right, we see a zoomed-in view of a portion of the arborescence

antonyms or opposites, insofar as inclusion does not allow inversion in the sense of opposition.

For instance, the animal–bird–sparrow series cannot be extended to the point of reversal into an opposite series. Another characteristic of modifiers is that they often refer to generic properties of a large number of nouns or verbs, either by nature when they express a permanent quality: "the immense sky" or by circumstance when they express a specific quality: "a blue flower". The two main axes of the space of adjectives in French computed by an FCA on the synonymy cliques are organized, respectively, between the normal–weird and good–bad oppositions. These major axes and all adjectives close to them can modify a very large number of lexies with a strong notional disparity between them. From these two properties (antonymy and great variability of what is being modified), it follows that the syntagmatic arborescences of adjectives used as epithets with an Adj-Nom pattern are very rich and diverse.

Here, I present two operators that work together to modify. The first is the epithet-adjective operator, which, to a given adjective, associates an arborescence composed of the lexies that it can modify. The second is the operator that associates to a lexie, usually a noun or a verb, the arborescence of its potential epithet adjectives, and for verbs, its potential adverbs.

9.2.7 Syntagmatic Arborescence of Adjectives

9.2.7.1 The Arborescence of Epithet Adjectives

Let m be a lexie, A_m its dual arborescence. Operator M is the operator that creates the syntagmatic arborescence A_m^{m-n} associated with m for the Adj-Nom pattern. Let's take the example of the French adjective *étroit* (narrow). It has a very broad scope and can just as well modify a person or a character as an object, or a place, etc. As a result, its syntagmatic arborescence (not shown here) is vast. The word *espace* (space) appears therein, as do a number of its synonyms, including *domaine* (domain), *intervalle* (interval), and *étendue* (extent). However, synonyms that are incompatible with modifying narrowness, like *immensity* or *infinite*, do not appear.

9.2.8 The Head-of-Modifiers *Operator*

Reciprocally, regardless of a lexeme's role in an actancial structure, it can invoke modifiers. Note in passing that I could very well have placed this paragraph in the ones about nouns or verbs. Let n be a noun and A_n its dual arborescence. Operator H is the operator that creates the arborescence A_n^{m-n} associated to n. For the word *espace* (space), although the syntagmatic arborescence A_n^{m-n} mainly contains modifiers that mark largeness like *considerable* (considerable), *grand* (big), *immense* (immense), *imposant* (imposing), *vaste* (vast), and *énorme* (enormous), there are also, albeit fewer in number, opposite modifiers such as *étroit* (narrow) and *petit* (small).

9.3 Composing Paradigmatic and Syntagmatic Arborescences Within a Sentence

As emphasized above, the results of neurophysiological and linguistic studies have pointed out the primacy of the verb. Here, arborescence composition follows the order of syntactic dependency, so the computation process starts with the verb. The specificity of this composition process, with respect to one that would follow a syntactic tree structure (like that of generative grammar or dependency grammar) is marked by the presence of the link induced by the verb between its arguments.

Figure 9.10 illustrates both the path followed when the composition process is applied to a sentence or phrase, and the underlying composition principle. For each branch of the syntactic tree, the paradigmatic and syntagmatic arborescences of its head and governed elements, and vice versa, are composed. Their product is the result of combining the syntagmatic arborescence of the appropriate head element (i.e., the one associated with a pattern that links the head and the governed element) and the paradigmatic arborescence of the governed element, as shown in Fig. 9.10, and vice versa. For example, the verb's syntagmatic arborescences will be connected, respectively, to the paradigmatic arborescences of its arguments. In return, an argument's syntagmatic arborescence will be connected to the verb's paradigmatic argument.

9.3.1 Composition Results of Our Illustrative Sentence

For the illustrative sentence used here, and for each node of the dependency tree, Fig. 9.11 shows the lexemes of the highest order extracted from the arborescences computed via the composition process. In addition to the shared elements, we can see some slight differences between the arborescences of the same word. For example, the word *espace* (space), when used in the phrase *cet étroit espace* (this narrow space), will have an arborescence derived from the composition that contains words that are consistent with this narrowness, such as *point* (point), *coin* (corner), and *place* (place). But when it is used in the phrase *(cet) espace renferme* ((this) space contains), the list of high-order words in the computed arborescence will contain *sphere* (area), *domain* (domain), and *étendue* (extent), which fit better with the larger size implied by the notion of containing something.

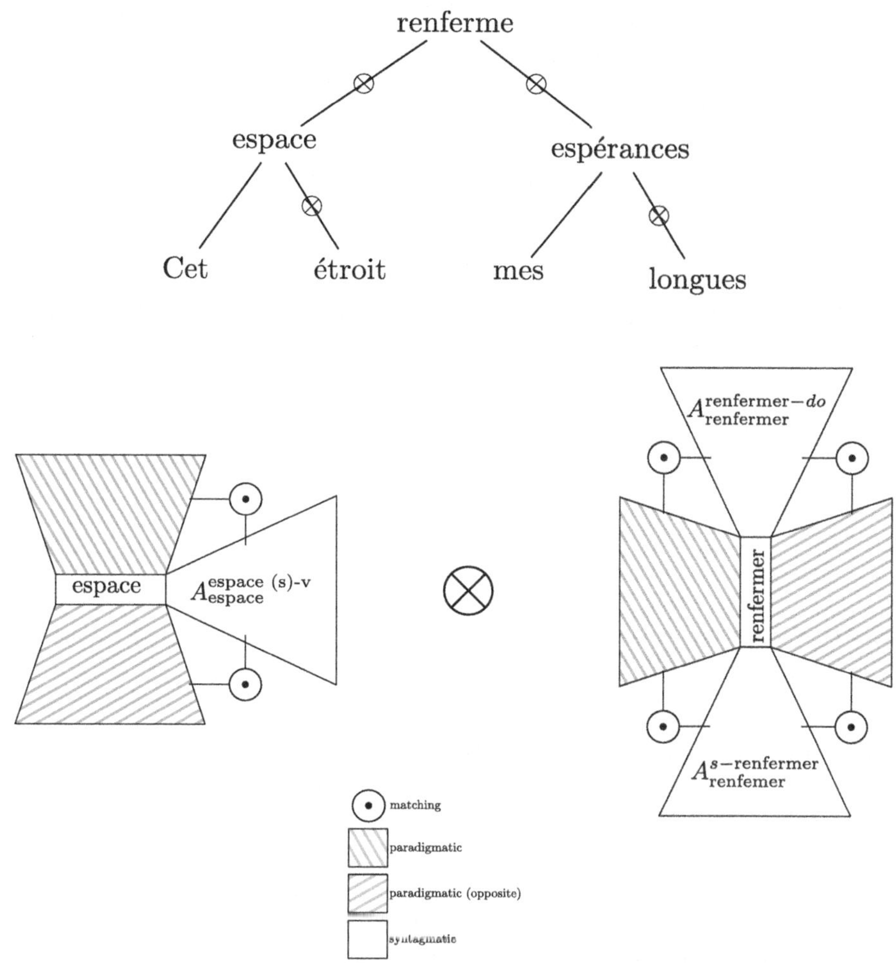

Fig. 9.10 Upper: A dependency tree. For each node in the tree, its dual paradigmatic arborescence is computed. Each edge corresponds to a specific "head element—governed element" pattern, in which both the head (h) and the governee (gve) are assigned their syntactic roles within the sentence. Then, at each node, the syntagmatic arborescences that result from this pattern are computed, following the method presented earlier. These arborescences are denoted A_h^{h-gve} and A_{gve}^{h-gve}. As indicated in the lower part of the figure, A_h^{h-gve} (respectively, A_{gve}^{h-gve}) is then combined with the paradigmatic arborescence A_{gve} of the *governee* (respectively, with the paradigmatic arborescence A_h of the head element). This combination is defined using the $\cap$ operator as follows: $C_h^{h-gve} = A_{gve}^{h-gve} \cap A_{h^p}$ if $A_{gve}^{h-gve} \cap A_{h^p}$ is not empty. If it is empty, $C_h^{h-gve} = A_{h^p}$. Likewise, $C_{gve}^{h-gve} = A_h^{h-gve} \cap A_{gve^p}$ if $A_h^{h-gve} \cap A_{gve^p}$ is not empty. If it is, then $C_{gve}^{h-gve} = A_{gve^p}$. The pair $(C_{gve}^{h-gve}, C_h^{h-gve})$ is matched using the relation induced by the syntactic pattern that connects the head element and the governed dependent. Finally, as stated above (Sect. 9.2.2), the arborescences computed for the verb's arguments are also matched using the relation induced by the verb-argument(s) pattern

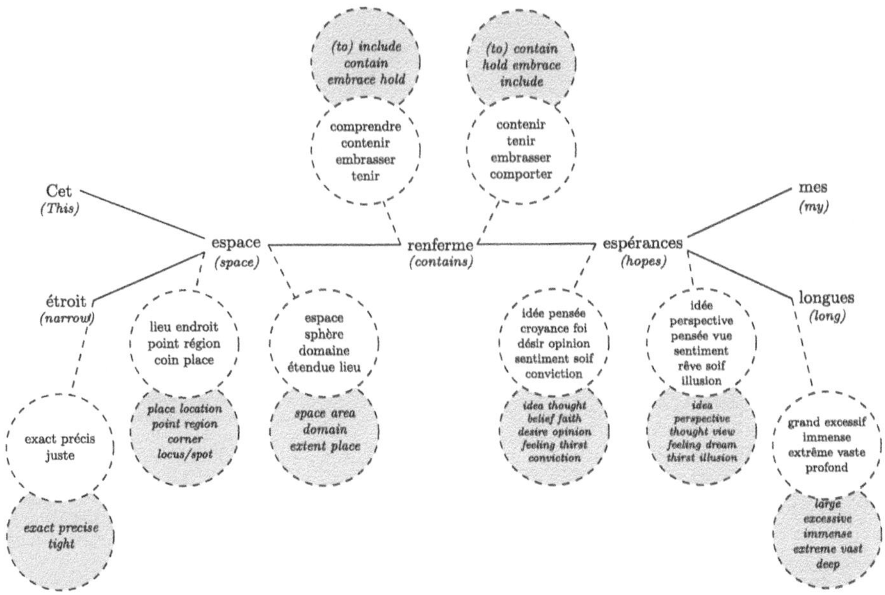

Fig. 9.11 For each lemma and each syntactic pattern to which it belongs, each circle contains the words of the highest order in the computed arborescence

9.4 Dual Arborescences and the Effects of Negation

Negation is expressed in a variety of ways (in French, using adverbs, such as *ne...pas* (not), *ne* (not), *ne...ni* (neither...nor), *ne...que* (nothing but). It can apply directly to lexies, as in *Il n'est ni grand, ni fort* (He is neither tall nor strong), in which case it has a proximal effect. We saw some examples of this in the preceding chapter. The scope of the negation can be more or less distal and also more or less indeterminate. The sentence *Mon cousin ne s'est pas fait renverser par une moto* (My cousin did not get run over by a motorcycle), an example inspired by Heldner, [6], can be paraphrased or made more explicit by "It's not my cousin who got run over by a motorcycle", or "My cousin got run over but not by a motorcycle", or "My cousin didn't get run over by a motorcycle (because he avoided it in time)". These different ways of paraphrasing arise from the scope of the negation within the sentence. A broader context than the one in the clause containing the negation can help resolve these ambiguities by constraining the possible interpretations.

Similarly, in the sentence *This étroit espace ne renferme pas mes longues espérances* (This narrow space does not contain my long-held hopes), and in the absence of a constraining context, the negation can apply to each of the lexemes of which it is composed. It can negate the verb (does not contain), the direct object (it contains something other than my long-held hopes such as my despair, my brief hopes, or someone else's long-held hopes, and so on). Note that from a semantic

Fig. 9.12 In each circle, the words of the highest order in the opposite arborescences of our illustrative sentence. Some sentences that could be derived from the combined computation of the composition in Fig. 9.11 and its opposite arborescences are *Cet étroit espace renferme mes profonds découragements* (This narrow space contains my deep discouragements), *Cet étroit espace révèle mes modestes espérances* (This narrow space reveals my modest hopes), *Ce bel espace renferme mes longues espérances* (This handsome space contains my long-held hopes…and thus, those potentially activated by a negation, depending on its scope

perspective, negation evokes opposites or antonyms. Thus, *ne pas renfermer* (to not contain) first suggests the opposite of *contain* before that of any other verb. The same holds true if the negation applies to *espérance* (hope), *long* (long-held), or *étroit* (narrow); it is from the paradigmatic space of the lexemes (and their opposites if there are any) that the possible paraphrases will be constructed. In the arborescence model, the negation operator inhibits the expressed part of the paradigmatic arborescences, and may even activate their opposite part. Figure 9.12 describes the potential effect of negation, and the paraphrases induced by the arborescences opposite to the ones computed during the composition process.

9.5 Semantic Interpolation (Continued)

In Sect. 8.5 of the previous chapter, I used dual paradigmatic arborescences to highlight semantic overlapping between words in the groups *étroit* (narrow); *long* (long-held) and *petit* (small); *espérance* (hope); and *illusion* (illusion), *ombre* (shade) and *soleil* (sun) from the excerpt of Mémoires d'outre-tombe.

Cet étroit espace me parut propre à renfermer mes longues espérances [...] Les arbres que j'y ai plantés prospèrent, ils sont encore si petits que je leur donne de l'ombre quand je me place entre eux et le soleil [...]. Un jour, en me rendant cette ombre, ils protégeront mes vieux ans comme j'ai protégé leur jeunesse [...] ils rappellent mes voyages et nourrissent au fond de mon cœur d'autres illusions.
This narrow space seemed suitable for containing my long-held hopes [...]. The trees I have planted here are thriving; they are still so small that I shade them when I stand between them and the sun [...]. One day, repaying me that shade, they will protect my old age as I protect their youth. [...] they remind me of my travels and, deep within my heart, they nourish other illusions. (our translation)

Now let's look at the example of the phrases *renferme mes espérances* (contains my hopes) and *j'ai protégé leur jeunesse* (I protected their youth) (in red) taken from the same excerpt. The paradigmatic arborescences of *renfermer* and *protéger* share the words *entourer* (to surround), *garder* (to keep), and *abriter* (to shelter); those of the lemmas *espérance* (hope) and *jeunesse* (youth) share the words *ardeur* (ardor) and *désir* (desire). Furthermore, the syntagmatic arborescences of the verbs *renfermer* (contain) and *protéger* (protect), when they have *espérances* (hopes) and *jeunesse* (youth) , respectively, as complement, both contain the words *force* (strength), *mouvement* (movement), *dessein* (intention), and *disposition* (frame of mind). These results demonstrate the ability of the arborescence-based model to make explicit a text's semantic interpolations (or isotopies), thereby enabling not only an analysis of how it unfolds and makes cross-references, even tenuous ones, but also, and consequently, its overall semantic cohesion.

9.6 Adaptability of the Arborescence Paradigm

I have presented a few composition phenomena for some simple syntactic structures. Other more complex structures or patterns could also be addressed, such as the syntax of function words. The mutual effects that these words have on the arborescences of their arguments can be represented in this manner. Moreover, the model could be adapted to transforming one form into another, such as going from the active to the passive voice and vice versa, or from a personal form to an impersonal form (e.g., Students were missing at role call $\Longrightarrow$ There were students missing at role call). It could also be extended to conversions, such as converting from a verb phrase to its nominalization (to reduce the rates $\Longrightarrow$ a reduction in rates) or converting from an adjective to a noun complement (e.g., She is intelligent $\Longrightarrow$ She is of great intelligence; a political matter $\Longrightarrow$ a matter of politics). In each case, the patterns connected by transformation or conversion can either be considered separately for the purpose of analyzing what is distinctive about each content form their arborescences attach to them, or collectively in order to determine what their respective content forms share.

References

1. Gosselin, L.: S mantique de la temporalit en fran ais. Un mod le calculatoire et cognitif du temps et de l'aspect. Recherches. Duculot, Louvain-la-neuve (1996)
2. Pasero, R., Royauté, J., Sabatier, P.: Sur la syntaxe et la sémantique des groupes nominaux à tête prédicative. Lingvisticæ Investigationes **27**(1), 83–124 (2004)
3. Madriñán, M.E.: IV. Les espaces collectifs de socialisation dans l'architecture de Rogelio Salmona. In: Cultures et créations dans les métropoles-monde, pp. 151–166. Hermann (2016)
4. Demanze, L.: Partager le paysage – sur Villa Zamir d'Hélène Gaudy. AOC media https://aoc.media/critique/2023/04/24/partager--le--paysage--sur--villa--zamir--dhelene--gaudy/ (2023)
5. Grevisse, M., Goosse, A.: Le Bon Usage. Duculot, De Boeck (2011)
6. Heldner, C.: La portée de la négation – Un examen de quelques facteurs sémantiques et textuels pertinents à sa détermination dans des énoncés authentiques. Ph.D. thesis, Universit de Stockholm (1981)
7. Thom, R.: Stabilité structurelle et morphogénèse. InterEditions, Paris (1977)

Conclusion

I have highlighted the constraints weighing on a system that links expression and content, the two heterogeneous constituents of language. Within the framework of a formal approach, this heterogeneity led me to account for language phenomena such as polysemy and the construction of meaning in context by proposing an arborescent format for the content form. This approach tends to mirror the semantic complexity of language.

Two issues are left open for future work. The first concerns the comparison with generative models of language, which currently dominate the field; the second pertains to the necessary consideration of content, an element already discussed in Sect. 9.1. I will briefly outline these two issues here.

Arborescent Models of Content Forms *vs* Vector-Based Models: Oppositions and Complementarities

As recalled earlier, models of language are vector-based, operate on very large corpora, and examine syntagmatic links only, without taking syntax into account. They have enabled the development of generative systems capable of producing texts that have been expressed in an astonishingly and increasingly fluent manner as these systems evolved. The fluidity of the texts produced is rooted in a probabilistic examination of the corpus on which they are based, and on the system architecture and algorithms implemented. By design, these models excel at producing texts that are similar or analogous to an input text and to texts from the source corpus. Here are some examples. In translation, the to-be-translated text and its translation are semantically similar; in summarizing, not only is the summary a condensed version of the source text but the source and its summary are semantically or structurally similar; in creative writing, the output text can be produced "à la manière de", thereby simulating an author, a literary genre, etc.

S. Ploux, *Semantic Matter and Its States*, Lecture Notes in Morphogenesis,
https://doi.org/10.1007/978-3-032-22809-3

Three aspects are proposed for comparing the various approaches. The first concerns the opposition between analysis and generation; the second compares the type of results obtained, i.e., the range of possibitilies vs the most probable response; the third pertains to the "cost" of the means employed.

Fine-Grained Analysis vs *Text Generation*

Because they are designed to find the masked word or the next word, or even the next sentence, vector models are best suited to production. When they are unsupervised, they are trained to give a probable textual response, and this training task is the basis of their ability to generate text in a given context. For translation, for example, the context consists of the sentence in the source language, the translation of the first few words, and expression regularities in the target language. These models have been criticized for their inability to understand their own output and sometimes for producing aberrant texts whose content may be erroneous.

Yet in humans, production and comprehension are distinct yet interdependent functions [1]. In terms of their brain location, these two functions rely on a large network of areas that are shared, as well as on networks and dynamics specific to each one [2, 3]. Comprehension is the result of a coherence that arises from a decomposing analysis and the reconstruction of the text's meaning.

Let us take metaphor as an example. Understanding a metaphor requires making the connection between its constituents and being able, as a result, to transfer semantic features from one to the other in order to construct an overall meaning. Thus, the sequence "money battle" is a metaphor based on the partial overlapping of verbs that are associated with each of the two nouns, e.g., *to win* and *to lose*, which govern both nouns (one can lose or win both money and a battle). Accordingly, syntagmatic arborescences are able to specify this verbal sharing onto which the notions of conflict between protagonists and ruthlessness in financial gain will be "grafted", thereby reinforcing understanding of the expression.

In contrast, a generative vector model will rely on analyses already present in its source corpus to reproduce something that is analogous. The analysis thus generated will have a relatively standard format that is developed across several paragraphs where the notions of financial conflict might occur.[1] However, the markers—such as shared associated verbs—supporting this analysis will not be computed or made explicit. The goal here is more to produce a generic analysis based on analyses existing in the corpus, but with a few different ingredients, than to detect what created cohesion in the expression.[2]

---

[1] This example is based on the inquiry "analyze the French expression bataille d'argent" input into ChatGPT3, May 2025.

[2] There are metaphor studies based on models of language, but most of them are detection systems. Few have attempted to analyze metaphors, and those that have done so tend not to look at connections between the metaphor's constituents, but to point out the domains they refer back to; the difference

Beyond the particular case of the processes underlying metaphors, the identification of isotopy—as proposed by Rastier [4] to support text analysis—cannot be taken into account by models that are aimed at learning how linguistic units are arranged in texts rather than at breaking down semantic units in a more detailed and deeper way in order to reveal the cohesion that emerges from their recomposition. The work presented here starts from the opposite side by proposing to model the content form.

However, just as language processes rely on the complementarity between comprehension and production, it seems legitimate—and certainly desirable—to have these two perspectives also complete each other, in order to take advantage of the duality of their approaches and the benefits likely to be provided by combining text analysis and text generation.

Moreover, the fine semantic granularity of the arborescent model could be incorproated into generative models in order to enhance them. Indeed, while models of languages have proven more effective when their input is not whole words but words broken down into tokens (frequent sequences of adjacent letters), the use of sub-units of meaning (the leaves of the arborescences) could also—as hypothesized here—contribute to their improvement.

Range of Possibilities vs *Most Probable Response*

Text generation in response to a prompt is an advantage that is highly appreciated by many users, even if they then rework the output to make it more complete and better aligned with their expectations. However, for those engaged in creative writing, it is more important to have tools that support the creative process than a ready-made response representing the most probable result with respect to the initial corpus. In this respect, synonym dictionaries have always been tools for obtaining good reformulations and also for finding a word one cannot remember. The arborescent paradigm extends this use by proposing enhanced possibilities that combine paradigmatic and syntagmatic relations.

Toward the Development of Parsimonious Models

Today's generative systems use a very large number of parameters (more than 10^{11} in some cases, compared to the number of entries in a dictionary, which ranges from roughly 5×10^4 to 5×10^5, depending on the language). These systems also require a massive training corpus. Clearly, the amount of energy consumed by their

between these domains is indicative of the presence of a metaphor (in our example here, the domains in question would be "war" and "money"). Furthermore, some of these systems rely on databases like WordNet to make use of the synonyms and hypernyms they contain, which amounts to reintroducing the paradigmatic relations initially discarded from language models.

computations is a source of increasing concern in the current context of climate change.

As I recalled in Chap. 7, these systems compute a single global space, whereas the approach proposed here computes arborescences for a word, a phrase, or a sentence on the basis of actual paradigmatic or syntagmatic relations. Computation of a representation specific to what is being processed and analyzed—independently of the entire lexicon and the language—considerably lightens the system. Lastly, one can reasonably predict that taking into account paradigmatic links (similarities and opposites) and syntactic patterns—both lacking in current language models based solely on textual regularities—would offer them an economy of resources whenever they ignore linguistic phenomena and attempt to compensate for this by scanning enormous training corpora.

Content Form and Content

A content form links expressions and contents. It is modelled here by systematizing differences between expression forms. However, the meaning of words, sentences, and texts also depends directly upon the content level. In the case of verbs, the content associated with a verb and the interaction it induces contain the outcome of the interaction. This outcome is directly accessible to any speaker or listener; neither needs an explanation to undestand that the sentence "The cat eats the mouse" ends the life of that mouse.

By essence, content is fluid, dynamic, and encompasses all aspects of cognition: sensory, motor, social interaction; it has access to the genesis and future of entities, of phylogenetic markers, etc. This mobility and this heterogeneity go beyond any transcription that could be made of them in expression forms and content forms, whenever the latter are constructed from the former.

I have proposed a formalism for representing and computing the internal, lexical-semantic structure that changes with the context of use, one that is not truly temporal, or dynamic, or spatial, or anchored in qualia.

Current approaches to the mathematization of temporality and interaction are dynamic (in the sense that they are dynamic systems like René Thom's model of verbal interaction), whereas models of spatiality are geometric. It would be necessary in any general formal theory of meaning to bridge the gap between approaches aimed at content-form modeling and those that start from the content itself. In this respect, I might mention neurogeometry [5, 6], which provides a mathematical grounding for the neuronal origin of spatial and perceptual representations. In this vein, one could imagine that each hyperedge of a content-form arborescence would be endowed with its own differential operator. The semantic unfolding of a language unit (word, phrase, sentence, or text) would show up as a solution in a space of embedded dynamics induced by all of these operators.

References

1. Pickering, M.J., Garrod, S.: An integrated theory of language production and comprehension. Behav. Brain Sci. **36**(4), 329–347 (2013)
2. Cai, J., Hadjinicolaou, A.E., Paulk, A.C., Soper, D.J., Xia, T., Wang, A.F., Rolston, J.D., Richardson, R.M., Williams, Z.M., Cash, S.S.: Natural language processing models reveal neural dynamics of human conversation. Nat. Commun. **16**(1), 3376 (2025)
3. Silbert, L.J., Honey, C.J., Simony, E., Poeppel, D., Hasson, U.: Coupled neural systems underlie the production and comprehension of naturalistic narrative speech. Proc. Natl. Acad. Sci. **111**(43), E4687–E4696 (2014)
4. Rastier, F.: Sémantique Interprétative. PUF, Paris (1987)
5. Petitot, J.: Elements of Neurogeometry. Springer, Berlin (2017)
6. Citti, G., Sarti, A.: Neurogeometry of perception: isotropic and anisotropic aspects. Axiomathes, 1–24 (2019)

The manufacturer's authorised representative in the EU is Springer Nature Customer Service Centre GmbH, Europaplatz 3, 69115 Heidelberg, Germany. If you have any concerns regarding our products, please contact ProductSafety@springernature.com

Printed and bound by CPI Group (UK) Ltd, Croydon, CR0 4YY
17/09/2026
02218438-0002